VOLUME ONE HUNDRED AND SEVENTEEN

ADVANCES IN PROTEIN CHEMISTRY AND STRUCTURAL BIOLOGY

Non-Heme Iron Enzymes – Structures and Mechanisms

AF902749

VOLUME ONE HUNDRED AND SEVENTEEN

ADVANCES IN PROTEIN CHEMISTRY AND STRUCTURAL BIOLOGY

Non-Heme Iron Enzymes – Structures and Mechanisms

Edited by

Tatyana Karabencheva-Christova
Department of Chemistry
Michigan Technological University
Houghton, MI, United States

Christo Z. Christov
Department of Chemistry
Michigan Technological University
Houghton, MI, United States

ELSEVIER

ACADEMIC PRESS
An imprint of Elsevier

Academic Press is an imprint of Elsevier
125 London Wall, London EC2Y 5AS, United Kingdom
525 B Street, Suite 1650, San Diego, CA 92101, United States
50 Hampshire Street, 5th Floor, Cambridge, MA 02139, United States
The Boulevard, Langford Lane, Kidlington, Oxford OX5 1GB, United Kingdom

First edition 2019

Copyright © 2019 Elsevier Inc. All rights reserved.

No part of this publication may be reproduced or transmitted in any form or by any means, electronic or mechanical, including photocopying, recording, or any information storage and retrieval system, without permission in writing from the publisher. Details on how to seek permission, further information about the Publisher's permissions policies and our arrangements with organizations such as the Copyright Clearance Center and the Copyright Licensing Agency, can be found at our website: www.elsevier.com/permissions.

This book and the individual contributions contained in it are protected under copyright by the Publisher (other than as may be noted herein).

Notices

Knowledge and best practice in this field are constantly changing. As new research and experience broaden our understanding, changes in research methods, professional practices, or medical treatment may become necessary.

Practitioners and researchers must always rely on their own experience and knowledge in evaluating and using any information, methods, compounds, or experiments described herein. In using such information or methods they should be mindful of their own safety and the safety of others, including parties for whom they have a professional responsibility.

To the fullest extent of the law, neither the Publisher nor the authors, contributors, or editors, assume any liability for any injury and/or damage to persons or property as a matter of products liability, negligence or otherwise, or from any use or operation of any methods, products, instructions, or ideas contained in the material herein.

ISBN: 978-0-12-818773-9
ISSN: 1876-1623

For information on all Academic Press publications visit our website at https://www.elsevier.com/books-and-journals

Publisher: Zoe Kruze
Acquisition Editor: Ashlie Jackman
Editorial Project Manager: Leticia M. Lima
Production Project Manager: Abdulla Sait
Cover Designer: Miles Hitchen

Typeset by TNQ Technologies

ELSEVIER Book Aid International
Working together to grow libraries in developing countries
www.elsevier.com • www.bookaid.org

CONTENTS

Contributors *vii*

1. Non-heme iron enzyme-catalyzed complex transformations: endoperoxidation, cyclopropanation, orthoester, oxidative C-C and C-S bond formation reactions in natural product biosynthesis **1**

Heng Song, Nathchar Naowarojna, Ronghai Cheng, Juan Lopez, and Pinghua Liu

1. Introduction 2
2. Verruculogen biosynthetic pathway 3
3. Fumitremorgin B oxidase (EAW25734) 24
4. Novofumigatonin biosynthetic pathway 27
5. EasH: an oxidase involved in cycloclavine biosynthesis 33
6. Orthoester formation in hygromycin B biosynthesis 36
7. Nogalamycin biosynthesis 39
8. Ergothioneine and ovothiol biosynthesis 42
9. Conclusion 49

Acknowledgments 49
References 49

2. Glutarate L-2-hydroxylase (CsiD/GlaH) is an archetype Fe(II)/2-oxoglutarate-dependent dioxygenase **63**

Caitlyn Q Herr, Lee Macomber, Efthalia Kalliri, and Robert P. Hausinger

1. Introduction to CsiD 64
2. Functional assignment of CsiD 67
3. Characterization of glutarate L-2-hydroxylase 73
4. Distribution of glutarate L-2-hydroxylase 80
5. Cellular role of glutarate L-2-hydroxylase 81
6. Conclusions 84
7. Experimental section 84

Acknowledgments 86
References 86

3. Comparison of DNA and RNA substrate effects on TET2 structure 91

Emmett M. Leddin and G. Andrés Cisneros

1. Introduction 92
2. Methodology 94
3. Results and discussion 96
4. Conclusions 109

Acknowledgments 109
References 109

4. Structure-function relationships in KDM7 histone demethylases 113

Shobhit S. Chaturvedi, Rajeev Ramanan, Sodiq O. Waheed, Tatyana G. Karabencheva-Christova, and Christo Z. Christov

1. Introduction 114
2. Biological and pathological role 115
3. Structure and mechanism 115
4. Dynamics of KDM7 enzyme-substrate complex 117
5. Reactivity studies of model system 121
6. Inhibitors of KDM7 122
7. Conclusions 123

References 124

CONTRIBUTORS

Shobhit S. Chaturvedi
Department of Chemistry, Michigan Technological University, Houghton, MI, United States

Ronghai Cheng
Department of Chemistry, Boston University, Boston, MA, United States

Christo Z. Christov
Department of Chemistry, Michigan Technological University, Houghton, MI, United States

G. Andrés Cisneros
Department of Chemistry, University of North Texas, Denton, TX, United States

Robert P. Hausinger
Department of Biochemistry and Molecular Biology, Michigan State University, East Lansing, MI, United States; Department of Microbiology and Molecular Genetics, Michigan State University, East Lansing, MI, United States

Caitlyn Q Herr
Department of Biochemistry and Molecular Biology, Michigan State University, East Lansing, MI, United States

Efthalia Kalliri
Department of Chemistry, Michigan State University, East Lansing, MI, United States

Tatyana G. Karabencheva-Christova
Department of Chemistry, Michigan Technological University, Houghton, MI, United States

Emmett M. Leddin
Department of Chemistry, University of North Texas, Denton, TX, United States

Pinghua Liu
Department of Chemistry, Boston University, Boston, MA, United States

Juan Lopez
Department of Chemistry, Boston University, Boston, MA, United States

Lee Macomber
Department of Microbiology and Molecular Genetics, Michigan State University, East Lansing, MI, United States

Nathchar Naowarojna
Department of Chemistry, Boston University, Boston, MA, United States

Rajeev Ramanan
Department of Chemistry, Michigan Technological University, Houghton, MI, United States

Heng Song
College of Chemistry and Molecular Sciences, Wuhan University, Hubei, People's Republic of China

Sodiq O. Waheed
Department of Chemistry, Michigan Technological University, Houghton, MI, United States

CHAPTER ONE

Non-heme iron enzyme-catalyzed complex transformations: endoperoxidation, cyclopropanation, orthoester, oxidative C-C and C-S bond formation reactions in natural product biosynthesis

Heng Song[a], Nathchar Naowarojna[b], Ronghai Cheng[b], Juan Lopez[b] and Pinghua Liu[b,*]
[a]College of Chemistry and Molecular Sciences, Wuhan University, Hubei, People's Republic of China
[b]Department of Chemistry, Boston University, Boston, MA, United States
*Corresponding author: E-mail: pinghua@bu.edu

Contents

1. Introduction 2
2. Verruculogen biosynthetic pathway 3
3. Fumitremorgin B oxidase (EAW25734) 24
4. Novofumigatonin biosynthetic pathway 27
5. EasH: an oxidase involved in cycloclavine biosynthesis 33
6. Orthoester formation in hygromycin B biosynthesis 36
7. Nogalamycin biosynthesis 39
8. Ergothioneine and ovothiol biosynthesis 42
9. Conclusion 49
Acknowledgments 49
References 49

Abstract

Non-heme iron enzymes catalyze a wide range of chemical transformations, serving as one of the key types of tailoring enzymes in the biosynthesis of natural products. Hydroxylation reaction is the most common type of reactions catalyzed by these enzymes and hydroxylation reactions have been extensively investigated mechanistically.

Advances in Protein Chemistry and Structural Biology, Volume 117
ISSN 1876-1623
https://doi.org/10.1016/bs.apcsb.2019.06.002
© 2019 Elsevier Inc.
All rights reserved.

However, the mechanistic details for other types of transformations remain largely unknown or unexplored. In this paper, we present some of the most recently discovered transformations, including endoperoxidation, orthoester formation, cyclopropanation, oxidative C-C and C-S bond formation reactions. In addition, many of them are multi-functional enzymes, which further complicate their mechanistic investigations. In this work, we summarize their biosynthetic pathways, with special emphasis on the mechanistic details available for these newly discovered enzymes.

1. Introduction

In natural product biosynthesis, in addition to their carbon skeleton assembly, modifications by tailoring enzymes play equally important roles in natural products' structural and functional diversities. Many genome-mining tools (e.g., antiSMASH (Blin et al., 2017), SMURF (Khaldi et al., 2010), ClusterFinder (Cimermancic et al., 2014)) are available to predict gene clusters by searching for enzymes involved in the construction of their carbon skeletons. After their carbon skeletons are assembled, in the biosynthesis of many natural products, tailoring enzymes then extensively modify the structures to add a second layer of complexities. Among these tailoring enzymes, non-heme iron enzymes are one of the largest enzyme superfamilies (Abu-Omar, Loaiza, & Hontzeas, 2005; Buongiorno & Straganz, 2013; Clifton et al., 2006; Costas, Mehn, Jensen, & Que, 2004; Hausinger, 2004; Hegg & Que, 1997; Kovaleva & Lipscomb, 2008; Kryatov, Rybak-Akimova, & Schindler, 2005; Martinez & Hausinger, 2015; Que & Ho, 1996; Solomon et al., 2000). They catalyze a wide range of reactions, including hydroxylation (Schofield and Hausinger, 2015), desaturation (Chang et al., 2014; Clifton et al., 2003), ring formation (Brauer, Beck, Hintermann, & Groll, 2016), ring expansion (Valegård et al., 1998), carbon skeleton re-arrangements (Matsuda, Awakawa, Wakimoto, & Abe, 2013), epoxidation (Li, van Belkum, & Vederas, 2012; Liu et al., 2001), epimerization (Chang et al., 2014; Clifton et al., 2003), and C-C bond cleavage (Blodgett et al., 2007). We have summarized many recent examples in a review paper last year (Gao, Naowarojna, Cheng, Liu, & Liu, 2018). In this paper, we focus on five more types of transformations reported recently: endoperoxidation, orthoester formation, cyclopropanation, oxidative C-C and C-S bond formation reactions. These transformations will be summarized in the context of their relevant biosynthetic pathways, with special emphasis on their proposed reaction mechanisms in conjunction with their available structural information.

2. Verruculogen biosynthetic pathway

Verruculogen (**1**, Fig. 1) is an endoperoxide containing terpene indole alkaloid isolated from several fungal strains, including *Penicillium verruculosum*, *Penicillia sp.* and *Aspergillus sp* (Williams, Stocking, & Sanz-Cervera, 2000; Fayos, Lokensgard, & Clardy, 1974). Structurally, it belongs to a class of alkaloids derived from L-tryptophan and L-proline with unique bicycle [2,2,2] ring system. Many of these compounds (Fig. 1) have been isolated and structurally characterized decades ago, while their biosynthetic pathways

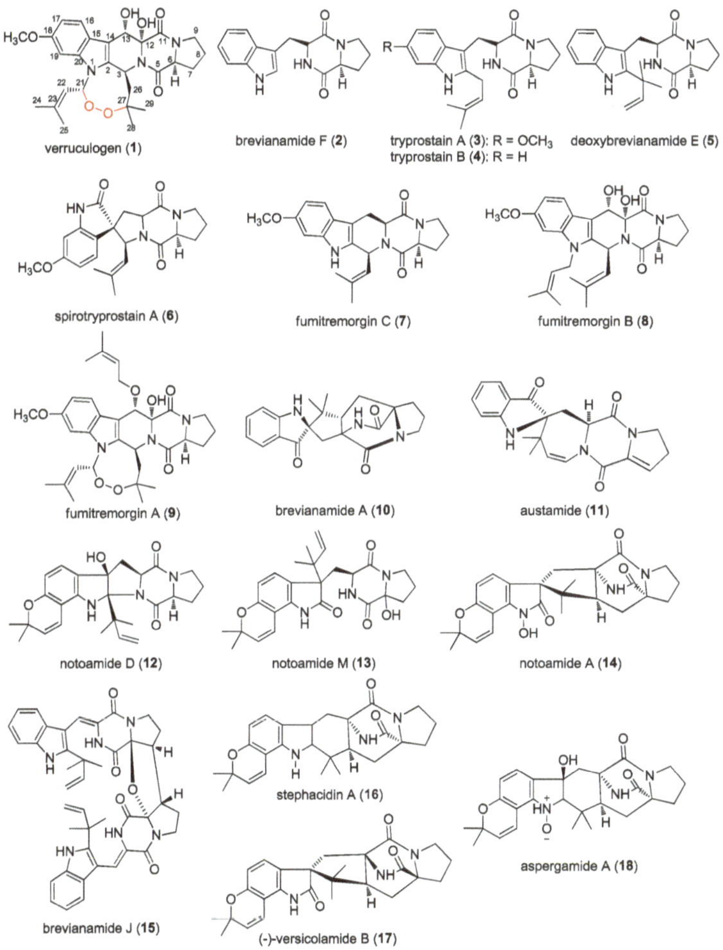

Fig. 1 Some representative examples of indole alkaloids with diketopiperazine skeleton, including verruculogen to be discussed.

started to be unraveled only in recent years, including verruculogen (**1**), tryprostatin A (**3**), spirotryprostatin A (**6**), fumitremorgins (**8**, **9**), brevianamide A (**10**), austamide (**11**) and notoamides (**12**–**14**) (Kato et al., 2007; Tsukamoto et al., 2008; Williams et al., 2000). This is a class of natural products rich in biological properties. For instance, verruclogen (**1**) is a tremorgenic mycotoxin that induces tremors in mice. It modifies the electrophysiological properties of human nasal epithelial cells (Khoufache et al., 2007). Several other compounds in this class are M-phase cell cycle inhibitors (Borthwick, 2012). Breast cancer resistance protein (BCRP) is a member of ATP-binding cassette family, which actively extrudes a wide range of structurally diverse molecules across cell's plasma membrane (Krishnamurthy & Schuetz, 2006), including drugs, toxins, endogenous compounds, and their metabolites. Fumitremorgin C (**7**) is an inhibitor of BCRP/ABCG2 (Rabindran, Ross, Doyle, Yang, & Greenberger, 2000) and could reverse the resistance of some tumor cell lines (Allen et al., 2002; Rabindran et al., 1998, 2000). Similarly, tryprostatin A (**3**) could inhibit the ATP-binding cassette transporters (ABC transporters) of various resistant cancer cell lines (Jain et al., 2008; Woehlecke, Osada, Herrmann, & Lage, 2003). Tryprostatins A (**3**) and B (**4**) as well as their diastereomers exhibit cytotoxicity toward many cancer cell lines (Zhao et al., 2002). Notoamide A (**14**) demonstrates moderate cytotoxicity against HeLa and L1210 cells (Kato et al., 2007). Given by the structural diversity of this class of natural products, the discovery of these biological activities opens up many opportunities for future systematic structure-function relationship studies.

Feeding studies using isotopically labeled precursors (both stable isotopes and radioactive isotopes) indicated that the verruculogen biosynthetic precursors are tryptophan (**19**), proline (**20**), dimethylallyl pyrophosphate (DMAPP), and methionine (Williams et al., 2000). Brevamide F (**2**), a synthetic cyclic dipeptide with a diketopiperazine structure, was efficiently incorporated into verruculogen. This result suggests that breviamide F (**2**) is probably one of the key verruculogen biosynthetic intermediates (Fig. 2) (Williams et al., 2000). In verruculogen, there are two prenyl groups, which are incorporated by prenyltransferases. For the most well characterized class of prenyltransferases (Bohlmann, Meyer-Gauen, & Croteau, 1998; Liang, Ko, & Wang, 2002; Sacchettini & Poulter, 1997), they usually contain one or more (N/D)DXXD sequence motifs, which are involved in prenyl diphosphate binding. These prenylation reactions depend strictly on the presence of divalent metal ions, e.g., Mg^{2+} or Mn^{2+} (Bohlmann et al.,

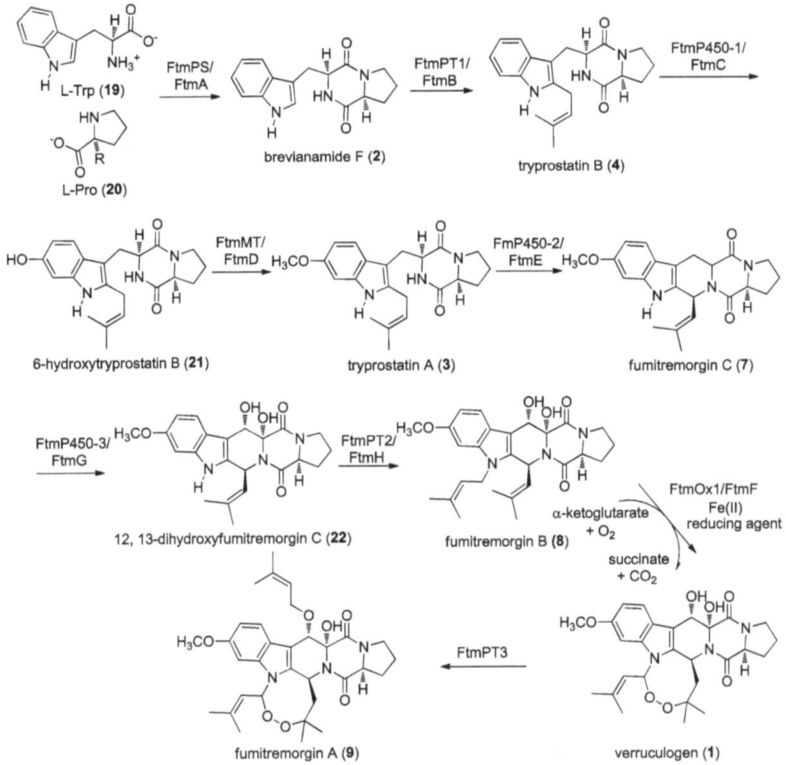

Fig. 2 Proposed fumitremorgin A biosynthetic pathway involving verruculogen as a biosynthetic intermediate.

1998; Liang et al., 2002). In recent years, another group of prenyltransferases were discovered (Pojer et al., 2003; Edwards & Gerwick, 2004). In the fungus *Claviceps* ergot alkaloid biosynthetic pathways, a dimethylallyltryptophan synthase (DMATS) catalyzes the prenylation of tryptophan at its C_4 position in a metal-independent manner (Tudzynski et al., 1999; Tsai, Wang, Gebler, Poulter, & Schardl, 1995). Sequences analysis reveals that DMATS does not contain the characteristic (N/D)DXXD motifs either. When Li and his co-workers analyzed the genome of one of the verruculogen producing strains, *Aspergillus fumigates*, they identified two DMATSs. One of them, FgaPT2, is involved in the biosynthesis of an ergot alkaloid, fumigaclavine C (Unsold & Li, 2005). The other one is located in a gene cluster containing nine putative biosynthetic genes (Grundmann & Li, 2005). Homologous gene clusters could also be identified in the genomes of some other putative verruculogen producing strains, including *Aspergillus fumigatus* A1163 and *Neosartorya fischeri* NRRL181 (Steffan, Grundmann, Afiyatullov, Ruan, & Li, 2009). The recombinant FtmPT1 could convert brevianamide F (**2**) to tryprostatin B (**4**, Fig. 2) in the presence of dimethylallyl diphosphate (DMAPP). Results from this study provide evidence indicated that the corresponding gene cluster is most likely the verruculogen biosynthetic gene cluster (Grundmann & Li, 2005). FtmPT1 could also be engineered to accept alternative substrates (e.g., hydroxynaphthalenes) as the prenyl group acceptors (Zhao et al., 2017).

The verruculogen producing strain, *A. fumigates* (Nierman et al., 2005), contains more than 20 gene clusters and several genes were predicted to be non-ribosomal peptide synthetases (NRPS). One of the predicted NRPS containing gene clusters also has the prenyltransferase (FtmPT1), which Li and co-workers have characterized. It was thus proposed that the bi-modular NRPS in this cluster (FtmPS) might be responsible for the activation and condensation of L-tryptophan (**19**) and L-proline (**20**) to form the diketopiperazine, brevianamide F (**2**). When Maiya et al. overexpressed FtmPS in a verruculogen producing strain, it led to a 2500-fold improvement in the production of brevianamide F (**2**) relative to its yield in the wild type strain (Maiya, Grundmann, Li, & Turner, 2006). In addition, when FtmPS was deleted, it abolished the production brevianamide F (**2**), fumitremorgin B (**8**) and fumitremorgin C (**7**, Fig. 2). These results support the assignment of this gene cluster as the verruculogen biosynthetic gene cluster.

After tryprostatin B (**4**) formation, the biosynthesis of verruculogen involves extensive modifications by tailoring enzymes. The *ftm* gene cluster has nine genes and three of them are predicted to be cytochrome P450s, including FtmC, FtmE, and FtmG. To assign the functions of these three cytochrome P450s, Kato et al. have characterized the metabolic profiles of their deletion strains using a fumitremorgin high production strain *A. fumigates* BM939 as the starting point (Kato et al., 2009). At the same time, they have also expressed these three P450 enzymes in yeast *Saccharomyces cerevisiae*. When metabolites from *ftmC* deletion strain was characterized, 6-hydroxyl-tryprostatin B (**21**) and tryprostatin A (**3**, Fig. 2) were absent, which highly suggested that FtmC was responsible for the hydroxylation of tryprostatin B (one of the two steps responsible for tryprostatin B **4** to tryprostatin A **3** transformation, Fig. 2). In the *ftmE* deletion strain, metabolites isolated did not have the C-N bond between the indole prenyl group and the diketopiperazine core, which implied that FtmE is responsible for the oxidative C-N bond formation to produce the fused rings in both verruculogen and fumitremorgin B (**3** → **7** transformation, Fig. 2). In FtmG deletion strain, fumitremorgin C (**7**) was detected, suggesting that both tryprophan hydroxylation and oxidative C-N bond formation in this strain were still functioning. Therefore, FtmG is most likely responsible for the dihydroxylation reaction of fumitremorgin C (**7**) to produce 12, 13-dihydroxyfumitremorgin C (**22**) (Fig. 2). When these three P450s were overexpressed in yeast, using a P450 reduction partner gene of *A. fumigatus*, *AFUA_2g07940*, Kato et al. have further demonstrated their *in vitro* biochemical activities using microsomes prepared from their yeast expression system (Kato et al., 2009). *In vitro*, FtmE did catalyze the oxidative C-N bond formation (**3** → **7** transformation, Fig. 2). In addition, FtmE can accept both tryptostatin A **3** and tryptostatin B **4** as the substrates for the oxidative C-N bond formation (Fig. 3). The mechanism for this oxidative cyclization reaction

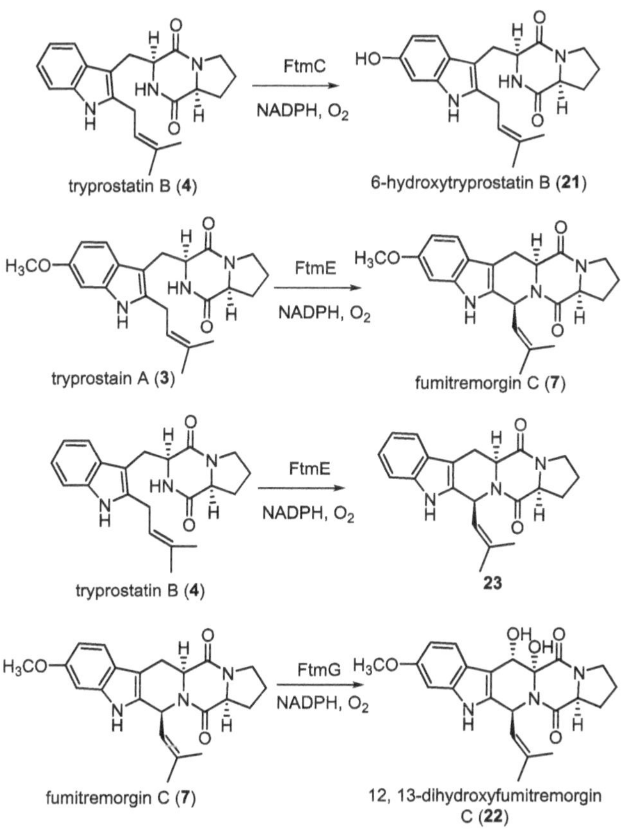

Fig. 3 Biochemical characterization of the three P450s in verruculogen biosynthetic pathway.

remains to be investigated. In *Streptomyces* sp. TP-A0274 staurosporine biosynthetic pathway, a bacterial cytochrome P450, StaN, catalyzes C-N bond formation between aglycon and deoxysugar moieties (Onaka, Asamizu, Igarashi, Yoshida, & Furumai, 2005). *In vitro*, dihydroxylation of fumitremorgin C (**7**) to 12, 13-dihydroxyfumitremorgin C (**22**) catalyzed by FtmG (Figs. 2 & Fig. 3) was indeed observed, while the *in vitro* activity was extremely low (Kato et al., 2009). Mechanistic details for FtmG-catalysis is still missing while results from feeding studies suggested that only one of the two hydroxyl group oxygens in 12, 13-dihydroxyfumitremorgin C (**22**) is derived from molecular oxygen (O_2) (Horak & Vleggaar, 1987).

To synthesize fumitremorgin A (**9**, Fig. 2), two more prenyltransferases are needed (Mundt, Wollinsky, Ruan, Zhu, & Li, 2012; Grundmann, Kuznetsova, Afiyatullov, & Li, 2008). In *A. fumigatus* Af293, the predicted verruculogen biosynthetic gene cluster contains two putative prenyltransferase genes (AFUA_8G00210 and AFUA_8G00250). One of them has been demonstrated to be the brevianamide F (**2**) prenyltransferase (FtmPT1, Fig. 2), which resulted in the production of tryprostatin B (**4**). In this gene cluster, there is one more prenyltransferase (FtmPT2). FtmPT2 shares ~30 % identity to FtmPT1. Grundmann et al. overexpressed and purified FtmPT2. Upon incubation of DMAPP and 12, 13-dihydroxyfumitremorgin C (**22**), fumitremorgin B (**8**) was observed (Kodato, Nakagawa, Hongu, Kawate, & Hino, 1988). These biochemical data thus provided unequivocal evidence supporting FtmPT2 as the second prenyltransferase (an *N*-prenyltransferase) in verruculogen biosynthesis, and its product is fumitremorgin B (**8**, Fig. 2). FtmPT2 has high substrate specificity in using DMAPP as a prenyl donor. No product was detected when geranyl diphosphate was used instead. However, FtmPT2 does show some degree of flexibility in terms of its prenyl acceptor, the aromatic substrate. Fumitremorgin C (**7**) was accepted by FtmPT2 with an activity of 10 % relative to that of 12,13-dihydroxyfumitremorgin C (**22**). Other biosynthetic precursors, such as brevianamide F (**2**), tryprostatin A (**3**) or B (**4**), were not accepted as a prenyl acceptor by FtmPT2. Similar to FtmPT1, FtmPT2 is Mg^{2+} or Ca^{2+}-independent and the presence of EDTA does not seem to affect its activity (Grundmann et al., 2008).

The *A. fumigatus* Af293 verruculogen biosynthetic gene cluster contains two prenyltransferase genes (FtmPT1 and FtmPT2). The production of fumitremorgin A (**9**) requires a third prenyltransferase, an *O*-prenyltransferase. In order to identify this gene, Mundt et al. analyzed a fumitremorgin A (**9**) producing strain, *N. fischeri* NRRL181 (Mundt et al., 2012; Nielsen,

Beuchat, & Frisvad, 1988) In *N. fischeri* genome, there exist ten DMATSs. Among these ten genes, seven of them have biological functions assigned already. Mundt et al. overexpressed the three remaining DMATSs and biochemically characterized them (Mundt et al., 2012). They demonstrated that NFIA_093400 was the verruculogen prenyltransferase, and named it FtmPT3. Mundt et al. also examined 73 potential aromatic substrates as potential prenyl acceptors for FtmPT3 catalysis, including L-tryptophan, L-tyrosine and their diverse derivatives, cyclic dipeptides of tryptophan and various amino acids, naringenin, (+)-catechin, 1-naphthol and 1,2-dihydroxynaphthalen, verruculogen (**1**), fumitremorgin B (**8**), and 12,13-dihydroxyfumitremorgin C (**22**, Fig. 2). Among these compounds tested, FtmPT3 could accept verruculogen (**1**), fumitremorgin B (**8**), or 12,13-dihydroxyfumitremorgin C (**22**) as the prenyl acceptor. When verruculogen (**1**) was used as the substrate, indeed, fumitremorgin A (**9**) was produced (Yamazaki, Fujimoto, & Kawasaki, 1980). Therefore, FtmPT3 is the third prenyltransferase, which installs the prenyl moiety to the aliphatic hydroxy group of verruculogen at the C_{13} position (**1** → **9**, Fig. 2). Similar to the other two prenyltransferases, the enzymatic activity of FtmPT3 was also divalent metal ion independent.

In verruculogen biosynthesis, the most intriguing transformation is the incorporation of the endoperoxide functionality (**8** → **1**, Fig. 2). Before the identification of the verruculogen endoperoxidase FtmOx1, it has also been suggested that the endoperoxide might be incorporated through spontaneous oxidation processes (Willingale, Perera, & Mantle, 1983). Hundreds of peroxy natural products have been isolated and structurally characterized (Casteel, 1992, 1999). The two most well-known examples are artemisinin (**23**, Fig. 4A) and prostaglandin G2 (**24**, Fig. 4B). The antimalarial agent artemisinin **23** from *Artemisia annua* is an endoperoxide-containing terpenoid (Zhao, Chang, Xiao, Liu, & Liu, 2013). Thus far, no information for its endoperoxide formation is available (Zhao et al., 2013; Paddon et al., 2013) while synthetic organic methods have been reported recently (Levesque & Seeberger, 2012). Prostaglandin G2 (**25**, PGG2) is part of the inflammatory signaling processes (Marnett, 2000). Before the mechanistic characterization of the verruculogen endoperoxidase, the only known example of endoperoxidase is prostaglandin H synthase (PGHS, Fig. 4B). PGHS is a heme-containing enzyme (Marnett, 2000; Smith, Urade, & Jakobsson, 2011; Stubbe & van Der Donk, 1998) and its catalytic process has two connected cycles (Fig. 4B): an activation cycle (shown in the up-left box of Fig. 4) and a catalytic cycle. In the activation cycle, PGHS

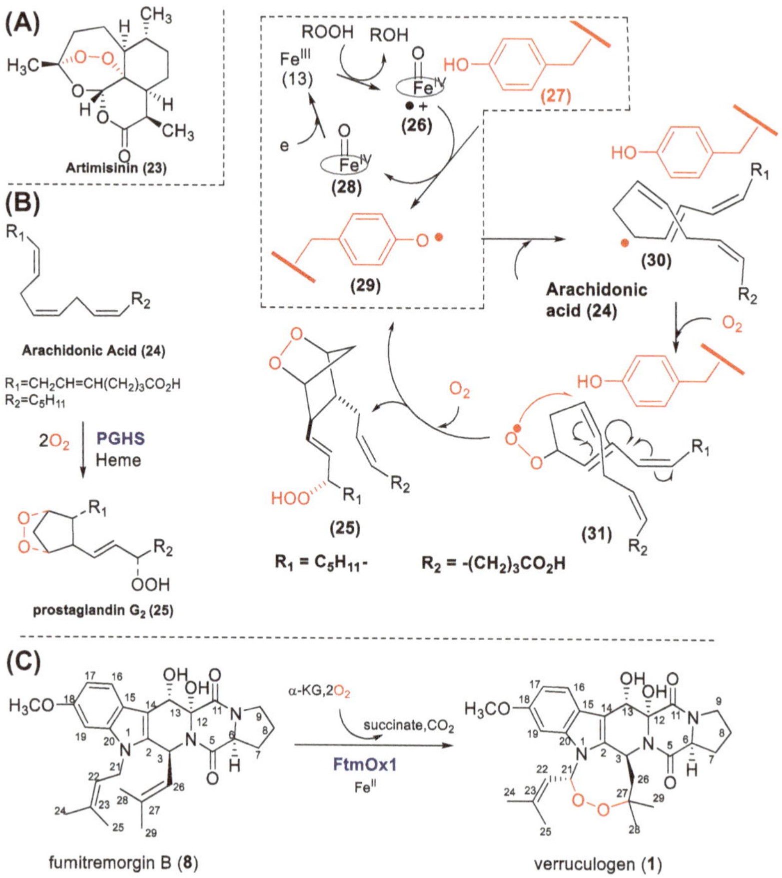

Fig. 4 Proposed mechanistic model for prostaglandin H synthase (PGHS). (A) structure of artemisinin; (B) Proposed PGHS mechanism; (C) FtmOx1-catalyzed endoperoxidation reaction.

heme center is oxidized by a peroxide to produce an $Fe^{IV}{=}O$/porphyrin cation radical (**26**, Fig. 4B), which then oxidizes an active site tyrosine **27** to a tyrosine-based radical **29**. Once tyrosyl radical **29** is produced, it can catalyze thousands of turnovers (**24** → **25**, Fig. 4B). Therefore, the heme center is only used for PGHS activation to create the active site tyrosine radical **29**, which is responsible for catalysis.

In the *A. fumigates* Af293 verruculogen biosynthetic gene cluster, it contains four potential oxygenases/oxidases. Three of them are cytochrome P450s (FtmC, FtmE and FtmG) and their functions have been assigned for the following three transformations: tryprostatin B (**4**) to 6-hydroxytryprostatin B (**21**, FtmC-catalysis), tryprostatin A (**3**) to fumitremorgin C (**7**, FtmE-catalysis), and fumitremorgin C (**7**) to 12,13-dihydroxyfumitremorgin C (**8**, FtmG-catalysis), respectively (Fig. 2). The two prenyltransferases and NRPS functions have also been assigned. To identify the gene responsible for the endoperoxidation reaction, there are only three genes in the gene cluster left with unknown functions, *ftmOx1* (also called *ftmF*), *ftmO* (*ftmI*) and *ftmMT* (*ftmD*). Based on sequence analysis, *ftmMT* (AFUA_8G00200) is methyltransferase and it is most likely responsible for the methylation of 6-hydroxytryprostatin B (**21**) to produce tryprostatin A (**3**) (Li, 2010). *ftmO* (AFUA_8G00260) is not related to verruculogen biosynthesis (Kato et al., 2009; Li, 2010). This left only one gene, *ftmOx1* (AFUA_8G00230), with no assigned function. Sequence analysis implies that FtmOx1 is an α-ketoglutarate-dependent dioxygenase. This is a surprising discovery as all α-ketoglutarate-dependent non-heme iron dependent dioxygenases activates molecular oxygen (O_2) with O-O bond cleavage to produce an Fe(IV)=O species for catalysis (Gao et al., 2018; Hausinger, 2004; Martinez and Hausinger, 2015). No endoperoxidation reactivity has previously been reported for this class of enzymes.

Using recombinant FtmOx1, Steffan et al. demonstrated that, in the presence of α-ketoglutarate and ascorbate, FtmOx1 could convert fumitremorgin B (**8**) to verruculogen (**1**, Fig. 4C) (Steffan, Afiyatullov, et al., 2009). At the same time, they detected two more side-products, although the identities of these two products were not assigned in their initial studies. However, this is the first demonstration of the presence of α-ketoglutarate-dependent non-heme iron dependent endoperoxidase. The formation of verruculogen (**1**) was strictly dependent on the presence of α-ketoglutarate. In addition, including EDTA (1 mM) to the reaction mixture completely abolished the production of verruculogen. None of the other metal ions, including Mn^{2+}, Co^{2+}, Cu^{2+} and Zn^{2+}, could replace Fe^{2+} for FtmOx1's endoperoxidation activity. These properties are all consistent with the properties of α-ketoglutarate dependent non-heme iron enzymes. Mechanistic studies on α-ketoglutarate dependent non-heme iron enzymes are one of the frontiers in enzymology and natural product biosynthetic studies because they catalyze transformations as wide as those heme-containing enzymes (Solomon et al., 2000; Gao et al., 2018). For α-ketoglutarate dependent non-heme iron enzymes, based on the initial model envisioned by Hanauske-Abel and Günzler (Hanauske-Abel & Günzler, 1982), a generic mechanism for this large class of enzymes was proposed and it involves

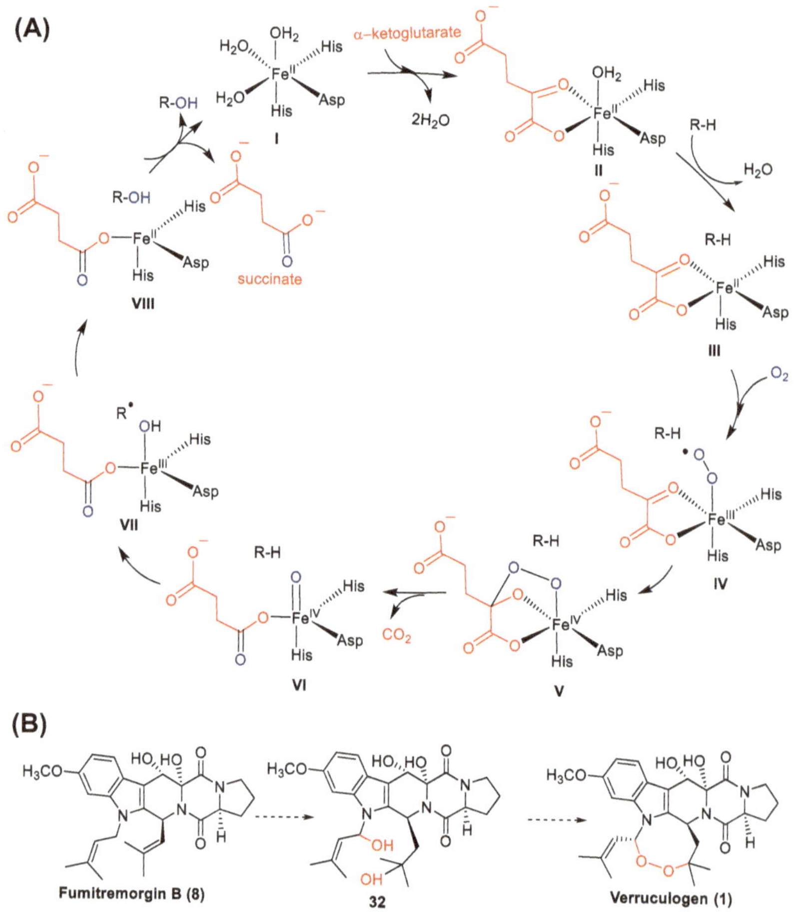

Fig. 5 A generic mechanism for α-ketoglutarate dependent non-heme iron enzyme catalyzed hydroxylation reaction. (A) The generic model. (B) Initial proposal for FtmOx1-catalysis, which involves two hydroxylation steps.

Fe(IV)=O species (**species VI**, Fig. 5A) as one of the key intermediates (Schofield and Hausinger, 2015; Huang & Groves, 2017; Kal & Que, 2017; Hangasky, Taabazuing, Valliere, & Knapp, 2013; Lundberg & Borowski, 2013; McDonough, Loenarz, Chowdhury, Clifton, & Schofield, 2010; Krebs, Galonic Fujimori, Walsh, & Bollinger, 2007; Solomon et al., 2000). Thus far, Fe(IV)=O intermediate (**species VI**, Fig. 5A) has been trapped and characterized from several enzymes already (Chang et al., 2014, 2016; Galonić, Barr, Walsh, Bollinger, & Krebs, 2007; Hoffart, Barr, Guyer, Bollinger, & Krebs, 2006; Matthews, Krest, et al., 2009,

Matthews, Nuamenn, et al., 2009; Price, Glass, et al., 2003, Price, Thirupathi, et al., 2003; Rohde et al., 2003). For all of the reactions characterized so far, it was suggested that Fe(IV)=O intermediate is the one responsible for the corresponding transformation (Gao et al., 2018), including hydroxylation, desaturation (Clifton et al., 2003), ring formation (Brauer et al., 2016), ring expansion (Valegård et al., 1998), halogenation (Weichold, Milbredt, & van Pée, 2016), and carbon skeleton re-arrangements (Matsuda et al., 2013).

Based on the generic mechanism for α-ketoglutarate dependent non-heme iron enzymes (Fig. 5A), the production of Fe(IV)=O intermediate involves O-O bond cleavage of molecular oxygen (O_2). To further examine this unprecedented endoperoxidation reaction catalyzed by FtmOx1, Steffan et al. have also conducted the activity assay in an $^{18}O_2$-enriched atmosphere and the product was then analyzed by mass spectrometry. Verruculogen (**1**) from this reaction contains an endoperoxide bridge containing either two [^{16}O]-atoms or two [^{18}O]-atoms. In this reaction, no product with a hybrid of [^{16}O]- and [^{18}O]-endoperoxide was detected. This result was also consistent with the conclusion from feeding experiments reported by Horak and Vleggaar (Horak & Vleggaar, 1987). Therefore, the possibility of verruculogen (**1**) formation by two typical hydroxylation steps (e.g., Fig. 5B) and subsequent oxidative cross-linking was excluded, because this model (Fig. 5B) predicts the formation of verruculogen (**1**) containing hybrid [^{16}O]- and [^{18}O]-containing endoperoxide under an $^{18}O_2$-enriched atmosphere. The same results have also been reproduced in our laboratory.

To account for these unique FtmOx1 properties, Steffan et al. has proposed a mechanistic model (Fig. 6) (Steffan, Yin, et al., 2009 Steffan, Afiyatullov, et al., 2009). In this model, the formation of the Fe(IV)=O species (**species E**, Fig. 6) from α-ketoglutarate dependent molecular oxygen activation follows a mechanism similar to other known α-ketoglutarate-dependent dioxygenases (Fig. 5A) (Gao et al., 2018; Hausinger, 2004) to produce the Fe(IV)=O species (**species E**, Fig. 6). The Fe(IV)=O species then abstracts a hydrogen atom from the fumitremorgin B C_{21} position to produce an allylic radical on the *N*-prenyl moiety (**species F**, Fig. 6). Similar to that in PGHS (Smith et al., 2011; Stubbe & van Der Donk, 1998), a second molecule of molecular oxygen (O_2) reacts with the *N*-prenyl moiety allylic radical, leading to the formation of peroxy radical (**species G**, Fig. 6). This peroxy radical then reacts with the other prenyl arm in fumitremorgin B (**8**) to form the endoperoxide-containing eight-membered ring. At the same time, a C_{26}-based radical (**species H**, Fig. 6) is formed.

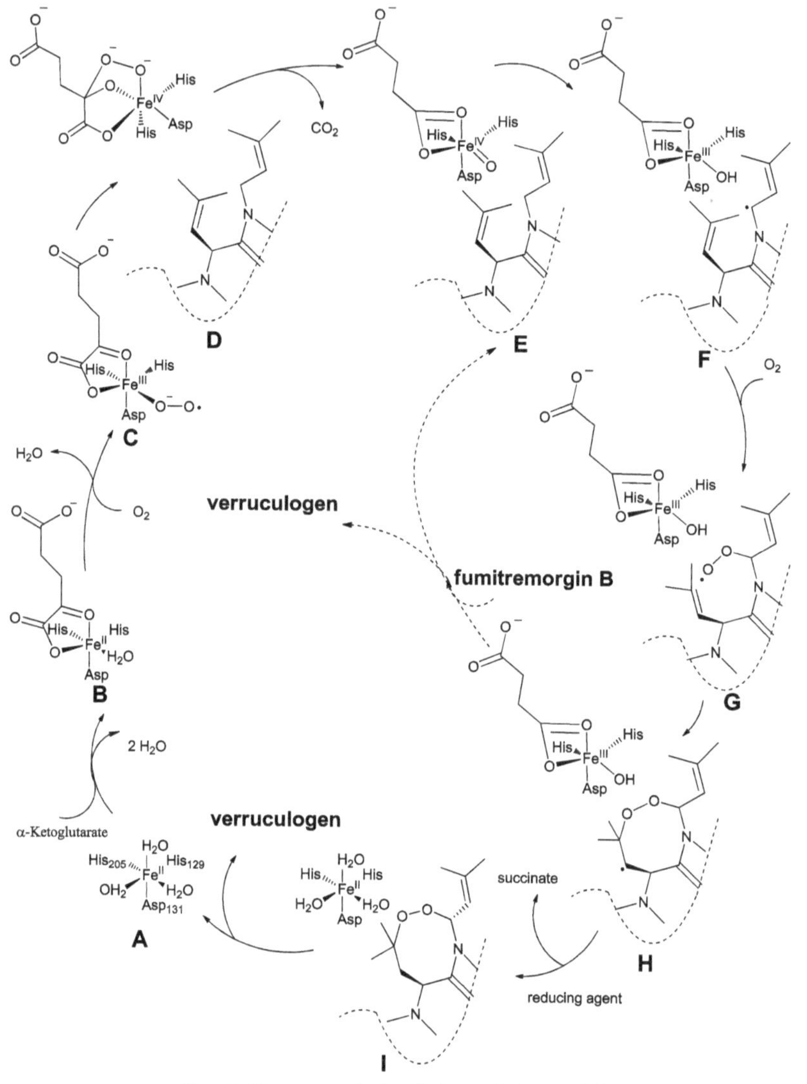

Fig. 6 The second FtmOx1-catalysis model.

To complete the reaction, the C_{26}-carbon based radical has to be quenched and it was suggested that the verruculogen radical (**species H**, Fig. 6) may abstract a hydrogen atom from free or enzyme-bound verruculogen. They also suggested that the C_{26}-based radical (**species H**, Fig. 6) may abstract one hydrogen atom from the Fe(III)−OH species in (**species H**, Fig. 6) to regenerate a Fe(IV)=O species (**species E**, Fig. 6), which would be used for the next catalytic cycle. The C_{26}-based carbon radical may also be reduced by ascorbate. To return the enzyme to its resting state, the Fe(III)-center also needs to be reduced to Fe(II). This mechanistic model raises two important questions. First, after the formation of the allylic radical on the *N*-prenyl arm in fumitremorgin B (**species F**, Fig. 6), what is the mechanism in FtmOx1 to prevent the typical hydroxy-rebound mechanism? Second, is ascorbate essential for this reaction?

The function of FtmOx1 as an endoperoxidase has also been confirmed by genetic studies. Kato et al. created the *ftmOx1* deletion strain from a fumitremorgin-producer stain, BM939, by replacing the entire coding region with the hygromycin B-resistance gene cassette (Kato et al., 2011). In the metabolic profiles of the *ftmOx1* deletion strain do not contain verruculogen (**1**), while the production of fumitremorgin B (**8**) was not affected. These genetic results are consistent with the assignment of FtmOx1 as the endoperoxidase by Steffan et al. (Steffan, Afiyatullov, et al., 2009). Built upon the studies from Steffan et al. (Steffan, Afiyatullov, et al., 2009), Kato et al. have also further characterized some of the unassigned products from FtmOx1-catalysis (Kato et al., 2011). One of these unknown compounds was assigned as 13-oxoverruculogen (**38**, Fig. 7) and the other one was assigned as 12α, 13α-dihydroxyfumitremorgin C (**35**, Fig. 7). Analysis of the mixture indicated that the amount of 12α, 13α-dihydroxyfumitremorgin C (**35**) produced is approximately 10% of the amount of verruculogen (**1**). The amount of 13-oxoverruculogen (**38**) is even less

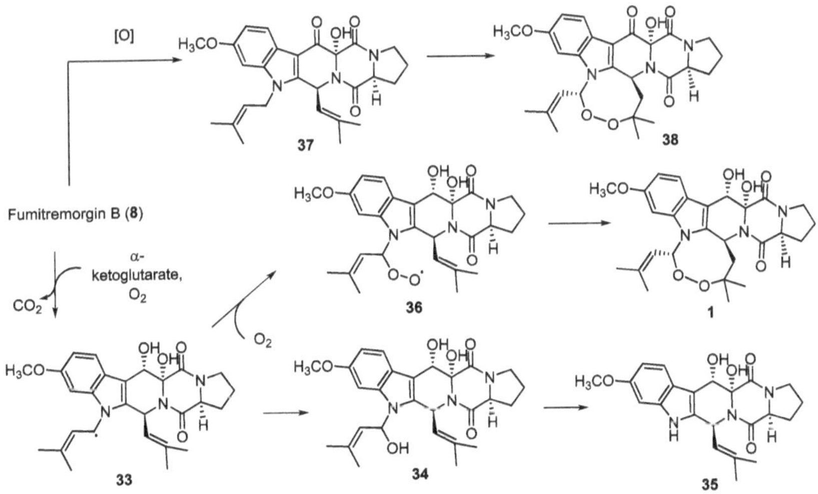

Fig. 7 Some side-reactions observed in FtmOx1-catalysis.

than that of 12α, 13α-dihydroxyfumitremorgin C (**35**). Kato et al. proposed that the production of 12α, 13α-dihydroxyfumitremorgin C (**35**) is most likely the decomposition product from an unstable hydroxylation product (**34**, Fig. 7). For the production of 13-oxoverruculogen (**38**), Kato et al. suggested that the oxidation of the C_{13}-hydroxyl group occurs first (**8** → **37**, Fig. 7), which is then followed by the endoperoxidation reaction (**37** → **38**, Fig. 7).

Before these biochemical characterizations, feeding studies using isotopically labeled mevalonolactone have also provided some key mechanistic information (Horak & Vleggaar, 1987; Vleggaar, Horak, & Maharaj, 1993). When (3*RS*, 5*S*)-[5-^{2}H, 4-^{13}C] mevalonolactone was fed into *P. verruculosum*, the resulting verruculogen (**1**) retains deuterium at its C_3-position. However, when (3*RS*, 5*R*)-[5-^{2}H, 4-^{13}C] mevalonolactone was feeding into *P. verruculosum*, the verruculogen C_{21} position retains deuterium. Results from these feeding studies suggested that the 5-R_e proton in mevalonolactone was retained in verruculogen C_{21} position, while the 5-*Si* proton in mevalonolactone was retained in verruculogen C_3 position (Horak & Vleggaar, 1987). When *P. verruculosum* was fed with [1-^{13}C, 2-^{2}H$_3$] acetate, deuterium was retained at both C_{22} and C_{26} position. Detailed NMR analysis led to the assignment of deuterium atom at the $C_{26}-R_e$ position (Fig. 8). One of the other important results from these feeding studies is the loss of stereochemistry at C_{27} position. When *P. verruculosum* was fed with (3*RS*)-[2-^{13}C] mevalonolactone, C_{24} position was labeled with [^{13}C] and no enrichment with [^{13}C] was observed at the C_{25} position. However, the results for C_{28} and C_{29} are quite different from that of C_{24}/C_{25} positions. At both C_{28} and C_{29} positions, there are significant levels of [^{13}C] enrichment, which highly suggests that from tryprostatin F (**2**) to verruculogen (**1**) biosynthetic process (Fig. 2), one of the steps can lead to a significant level of bond-rotation around C_{26} and C_{27} positions (Fig. 8). These feeding study

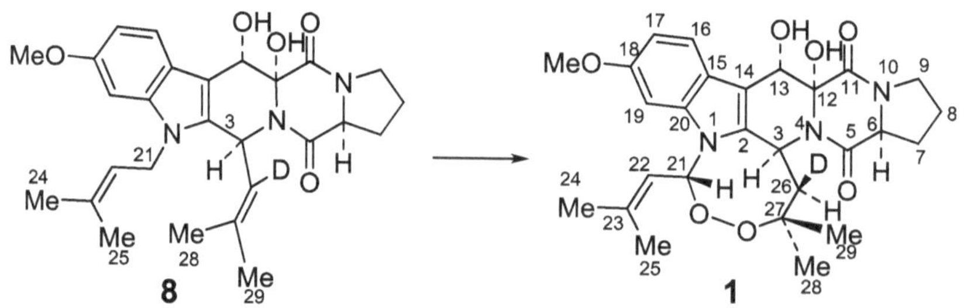

Fig. 8 Results from verruculogen feeding studies.

results have to be considered when postulating mechanism of these enzymes involved in verruculogen biosynthesis.

To mechanistically characterize this unprecedented endoperoxidation reaction, by a combination of biochemical and biophysical approaches, we analyzed this endoperoxidation reaction by addressing whether this reaction requires any external electrons (Yan et al., 2015). Interestingly, under single turnover conditions, reductant was not essential for FtmOx1-catalysis. Therefore, the ascorbate discussed in the initial report by Steffan et al. is not required for FtmOx1-catalysis under single turnover conditions. In addition, it was discovered that under single-turnover conditions, the product profile was also very different from the early reports (Kato et al., 2011; Steffan, Afiyatullov, et al., 2009). Under single turnover conditions and in the absence of any reductants, Fe^{2+}-reconstituted FtmOx1 produced two products, verruculogen (**1**) and 13-oxoverruculogen (**38**), having 13-oxo-verruculogen (**38**) as the dominant product (Fig. 7). In addition, under single-turnover conditions, the stoichiometries of each of the components in FtmOx1-catalysis were analyzed. Under single-turnover conditions, each turnover consumes one equivalent of α-ketoglutarate and two equivalents of molecular oxygen (O_2).

We have also obtained FtmOx1 crystal structures including FtmOx1 alone and its binary complex with α-ketoglutarate. FtmOx1 exhibits a typical a double-stranded β-helical fold (DSBH fold) DSBH fold, which is common for α-ketoglutarate dependent non-heme iron enzymes. The Fe^{2+} center is ligated by a conserved 2-His-1-carboxylate facial triad (H129, D131, and H205, Fig. 9B). The other three sites are occupied by waters. In the FtmOx1·Fe^{2+}·α-ketoglutarate binary complex, α-ketoglutarate binds to the Fe^{2+}-center bi-dentately and an additional water molecule occupies the remaining coordination site (Fig. 9B). In the α-ketoglutarate dependent non-heme iron enzymes characterized so far, two different α-ketoglutarate binding configurations have been observed, the proximal (TauD in Fig. 9A) and distal (FtmOx1 in Fig. 9B) types (Gao et al., 2018; Hangasky et al., 2013; Clifton et al., 2006). In the proximal type α-ketoglutarate binding mode (e.g., TauD·Fe^{2+}·α-ketoglutarate binary complex in Fig. 9A), the α-ketoglutarate C_1 carboxylate coordinates to the Fe^{2+} center *trans* to the first histidine (His99). In FtmOx1·Fe^{2+}·α-ketoglutarate binary complex shown in Fig. 9B, the α-ketoglutarate C_1 carboxylate is at a position opposite to the distal histidine (His205) of the 2-His-1-carboxylate facial triad (Fig. 9B) (Yan et al., 2015). Thus far, these two types of α-ketoglutarate binding configurations exist in roughly 1:1 ratio among reported

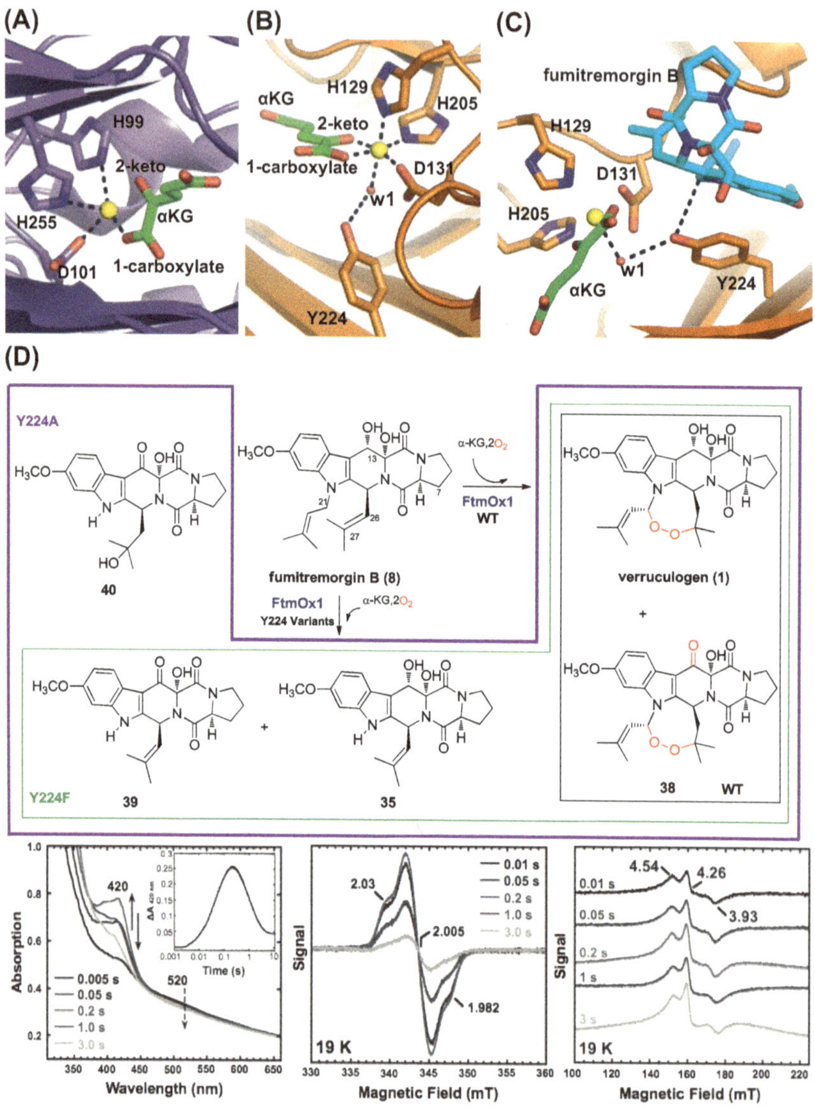

Fig. 9 FtmOx1 crystal structural information. (A) TauD structure; (B) Crystal structure of FtmOx1·Fe^{2+}·α-ketoglutarate binary complex; (C) Model of FtmOx1·Fe^{2+}·α-ketoglutarate·fumitremorgin B tertiary complex; (D) Reaction products in FtmOx1 Y224F and Y224A catalysis and pre-steady state and spectroscopic characterization of some intermediates in wild-type FtmOx1-catalysis.

structures of α-ketoglutarate non-heme iron enzymes in the protein data bank (PDB). In the FtmOx1·Fe^{2+}·α-ketoglutarate binary complex, it was also discovered that Tyr224 is right next to the Fe^{2+}-center and forms a hydrogen bond with one of the Fe^{2+}-center water ligand.

Despite extensive efforts, we have not yet obtained the crystal structure of the FtmOx1·Fe^{2+}·α-ketoglutarate·substrate tertiary complex. With more than 30 datasets obtained from either soaking with substrate fumitremorgin B (**8**) or co-crystallization with fumitremorgin B (**8**), consistent electron densities near the iron center were observed. These electron densities were not sufficient to model the whole fumitremorgin B (**8**) molecule and could be alternatively interpreted as the buffer molecules along with a chain of water molecules. However, we proposed that this is most likely the substrate binding site after several other factors were taken into consideration. As fumitremorgin B (**8**) is a relatively large molecule, searching of other potential binding site using DogSite did not identify other pocket that is large enough to accommodate fumitremorgin B (**8**) near the iron center. Furthermore, in our X-ray crystallographic studies, with more than 30 data sets, we consistently observed additional connected densities in this hydrophobic pocket even when buffers were extensively washed-off. We have also analyzed the electron densities using the Phenix LigandFit program to see whether this is a potential substrate binding pocket. Due to the large size and unique shape of the substrate, there are only two orientations in which the substrate could fit into the density. Eventually we modeled the substrate into this electron density with ~60% occupancy. The challenges faced in obtaining high occupancy FtmOx1·Fe^{2+}·substrate binary complex could be explained by several challenges faced in FtmOx1 studies. First, fumitremorgin B (**8**) is a large hydrophobic molecule. It is difficult to obtain high concentrations of fumitremorgin B (**8**) in the aqueous buffer system. Second, the potential substrate binding pocket we used to model fumitremorgin B (**8**) is hydrophobic. In this hydrophobic environment, the substrate may bind with some degree of conformational flexibility. Third, as we will discuss later (Fig. 10), FtmOx1-catalysis involves two different reactions (endoperoxidation and the oxidation of the C_{13} hydroxyl group to a keto), therefore, FtmOx1 catalysis involves at least four different positions on a large molecule. A high degree of flexibility of fumitremorgin B (**8**) in the FtmOx1 active site is probably needed to accomplish these transformations. Taking all these factors into consideration, we proposed that the site where extra electron densities were observed is the fumitremorgin B (**8**) binding site. We modeled fumitremorgin B (**8**) into these observed electron

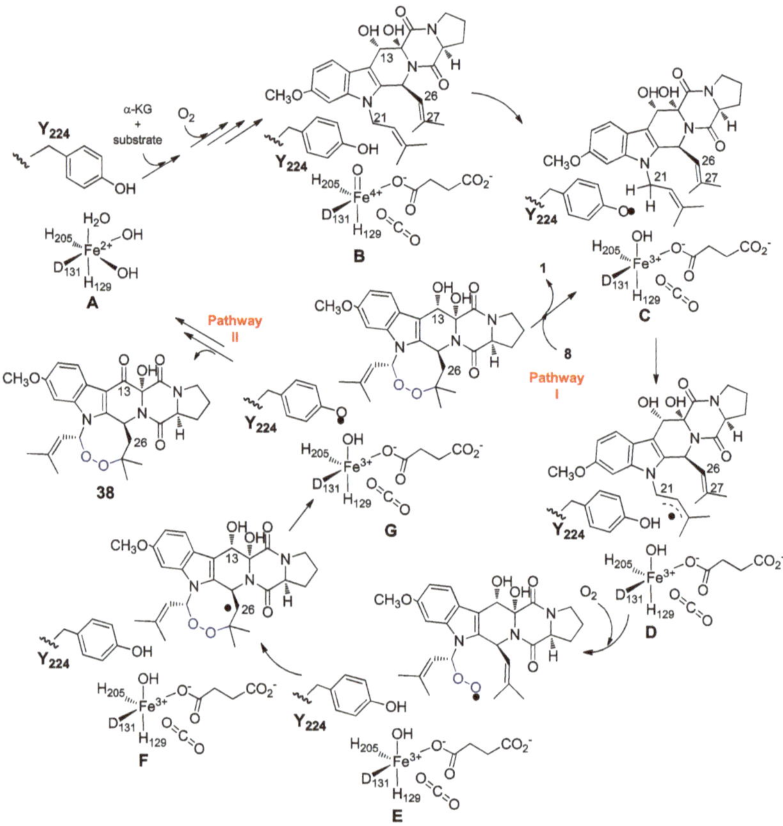

Fig. 10 Our FtmOx1 mechanistic model involving Y224-based tyrosyl radical.

densities in the hydrophobic pocket with ~60% occupancy. In this model (Fig. 9C), rings A and B of the substrate, fumitremorgin B (**8**), form π-π stacking with Tyr224. Overlaying the iron centers in these two structures (FtmOx1·Fe^{2+}·α-ketoglutarate binary complex and FtmOx1·Fe^{2+}·substrate binary complex) allows us to visualize the relative positions of the Fe^{2+} center, Tyr224, and the substrate. The distal-type of α-ketoglutarate binding configuration creates a situation where fumitremorgin B (**8**) and the metallo-center are separated by Tyr224, which immediately raises the question as how the endoperoxidation reaction occurs. Similar to other α-ketoglutarate dependent non-heme iron enzymes, the FtmOx1·α-ketoglutarate binary complex is pink in color with an absorption feature centered at ~520 nm, which originates from a typical metal-ligand charge transfer. In the absence of substrate fumitremorgin B (**8**), exposing the FtmOx1·Fe^{2+}·α-ketoglutarate binary complex to molecular oxygen (O_2)

leads to the formation of a blue colored species with absorption centered at ~ 600 nm (ε_{600} ~ 380 M^{-1} cm^{-1}). Tandem mass spectrometry analysis of this species indicated the formation of the dihydroxyphenylalanine (DOPA) from a tyrosine residue (Tyr224), which is consistent with the structural information, suggestive of Tyr224 proximity to the iron center. This FtmOx1 self-hydroxylation kinctic information matches with other examples in literature (Liu, Ho, et al., 2001, 2004; Ryle et al., 2003). In TauD, a tyrosyl radical can be observed at ~20 s time-scale and there is very slow decay of the tyrosyl radical, leading to DOPA formation (30 min time-scale). In FtmOx1, the catalytic cycle is completed within a second, while this self-hydroxylation is on the minute time-scale and DOPA formation takes up to 30 min. To test the role of Tyr224 in FtmOx1-catalysis, we created Y224F and Y224A mutants and characterized them biochemically. In both mutants, there is a small amount of verruculogen (**1**) formation, while the dominant products are from the classical hydroxylation/dealkylation type of reactions (Fig. 9D), leading to the production of **35**, **39**, or **40** as the dominant products.

FtmOx1-catalysis has also been characterized by stopped-flow optical absorption spectroscopy (SF-Abs) and freeze quench (FQ) coupled electron paramagnetic resonance spectroscopy (EPR). A transient radical intermediate was observed exhibiting an optical absorption feature centered at ~420 nm. This species reached a maximum at ~0.2 s and decayed within a second. At a longer time frame (~10 s), another radical species was observed. However, this radical species is accumulated to a much lower level relative to the transient species at ~0.2 s. In addition, this ~10 s species is not kinetically competent because the enzyme completes one turnover within a second. This transient species intermediate ~0.2 s were further characterized by FQ-EPR experiments (Fig. 9E). after the rapid mixing of O_2-saturated buffer with the FtmOx1·Fe^{2+}·α-ketoglutarate-fumitremorgin B complex, two EPR signals were observed and they maximized at ~0.2 s. The first set of EPR signals (g = 4.54, 4.26, and 3.93) belong to a high-spin Fe^{3+} species having axial and rhombic zero-field splitting parameters of $|D| < 0.5$ cm^{-1} and $E/D \approx 0.26$, respectively. The second EPR signal in the $g = 2$ region is most likely from a radical species. The formation and decay of this radical signal matches with the kinetics of the SF-Abs 420 nm absorption feature. Therefore, we proposed that they are from the same intermediate. The width of the radical EPR signal (~12 mT edge-to-edge width) was significantly broader than that of magnetically isolated organic or protein radical signals (Stubbe & van Der Donk, 1998), which could be due to magnetic dipolar interactions between the radical species and the Fe^{3+} center.

We proposed a mechanism to account for all of these results (Fig. 10). In this mechanism, several radical intermediates are involved: tyrosyl radical (species **C** and **G**), substrate radical (species **D**), peroxyl radical (species **E**) and product radical (species **F**, Fig. 10). One important aspect of this FtmOx1 mechanistic model is the function of Tyr224. The current hypothesis suggests that Y224 may be oxidized by the ferryl intermediate to form a tyrosyl radical, which is the actual key oxidant to perform the initial C-H activation (species **C** → species **D**, Fig. 10). A recent quantum chemical study also supports this hypothesis (Jakubczyk et al., 2015). In addition, this hypothesis is consistent with biochemical results in the Y224F variants, which could still oxidize the substrate C-H bond through the direct hydroxylation pathway, to afford the dealkylated product as the dominant product. Therefore, the strategic positioning of Y224 in FtmOx1 is probably used to block the typical hydroxyl rebound mechanism so that the dominant reaction will be endoperoxidation.

Recently, an alternative mechanistic model of FtmOx1 has been proposed, suggesting that either α-ketoglutarate or Fe(IV)=O formed in FtmOx1 may rotate from the current position (Fig. 9B) to face fumitremorgin B (**8**) directly. This rearrangement allows the Fe(IV)=O species to abstract a hydrogen from fumitremorgin B (**8**) C_{21} position directly. However, this hypothesis did not take into account of one important detail. In our single turnover experiment, there is no other reductant (e.g., no ascorbate) and the dominant product is 13-oxoverruculogen (**43**). In order to explain the production of 13-oxoverruculogen (**43**) as the dominant product, a potential ferryl regeneration model (the regeneration of Fe(IV)=O species from the C_{26}-based organic radical, **species D → species E**, Fig. 11) was suggested. This hypothesis is highly unlikely based on the information in literature from both experimental and theoretical studies (de Visser, 2010). For the proposed carbon-entered radical to abstract a hydrogen atom from Fe^{3+}—O—H to regenerate Fe(IV)=O species, it must overcome an activation barrier of over 20 kcal mol^{-1}. In contrast, the competing hydroxyl rebound pathway has an activation barrier of < 3 kcal mol^{-1} (or effectively barrierless in some cases). Therefore, it is unlikely that the ferryl regeneration model (**species D → species E**, Fig. 11) will be feasible in FtmOx1-catalysis.

Another model (a CarC-type model) has also been suggested by some scientists at an enzyme mechanism conference. CarC was proposed to

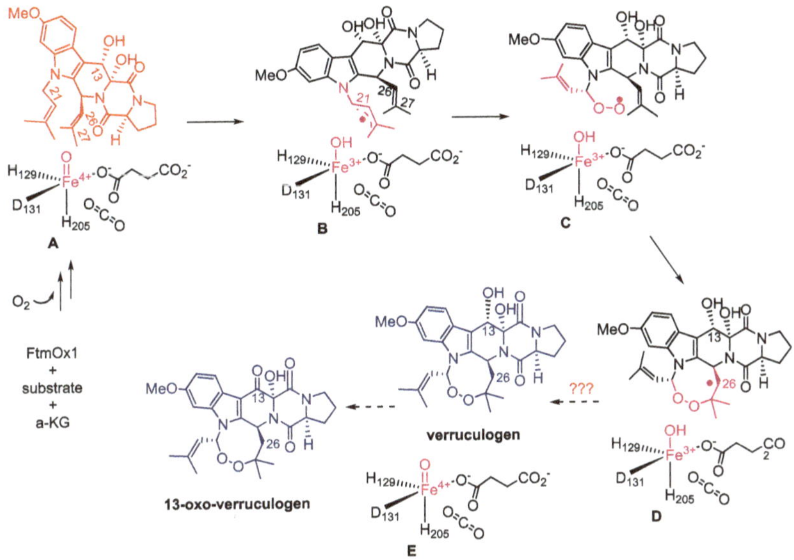

Fig. 11 Ferryl regeneration model for FtmOx1 catalysis proposed by some scientists.

catalyze both epimerization and dehydrogenation reactions. However, CarC-catalysis is achieved through two cycles of α-ketoglutarate based O_2 activations. In addition, a reductant is needed after the epimerization cycle to return the enzyme to its Fe^{2+} state. In our single turnover studies of FtmOx1, the production of 13-oxoverruculogen (**38**) as the dominant product does not require a reductant such as ascorbate. Therefore, both endoperoxidation and oxidation of C_{13}-hydroxyl group were achieved after one cycle of α-ketoglutarate oxidative decarboxylation based O_2 activation. As a result, FtmOx1-catalysis is fundamentally different from CarC-catalysis (Chang et al., 2014; Phelan & Townsend, 2013). In addition, this model suggests that Y68 might play a key role in this CarC-type of model. With respect to our unpublished results, the Y68F mutant still produces 13-oxo-verruculogen (**38**) as the dominant product under our single turnover conditions. Therefore, this CarC-type of model is not consistent with current experimental results either.

Recently, FtmOx1-catalyzed endoperoxidation has also been investigated by two computational studies (Wang, Su, & Liu, 2017; Ji & Chen, 2018). In the first study reported by Wang et al. (2017), the overall mechanistic model outlined in Fig. 10 was examined by a combination of quantum mechanics and molecular mechanics (QM/MM) method. A few key features of the model in Fig. 10 were examined. First of all, after the formation of the Fe(IV)=O species, what are the activation barriers for the oxidation of Tyr224, and the activation barrier for the fumitremorgin B (**8**) C_{21} hydrogen abstraction by Tyr224 based radical? Their calculation indicated that the Fe(IV)=O species abstracts a hydrogen from the hydroxyl of Tyr224 to produce a Tyr224-based radical with an energy barrier of 9.1 kcal mol^{-1}. In the subsequent reaction, the Tyr224 radical abstracts a hydrogen atom from fumitremorgin B (**8**) C_{21} position to produce a substrate-based radical with an energy barrier of 23.8 kcal mol^{-1}. For the alternative pathway, direct hydrogen atom abstraction by Fe(IV)=O at fumitremorgin B (**8**) C_{21} position has an activation barrier of 33.9 kcal mol^{-1}, which are consistent with the model outlined in Fig. 10 (Yan et al., 2015). Thus far, their calculation did not lead to any reasonable explanation for the second cycle of FtmOx1-catalysis: the oxidation of C_{13}-hydroxyl group to 13-oxoverruculogen (**45**). Therefore, many more studies will still be needed for this unprecedented transformation. Recently, Chen and co-workers have reported another key information on the FtmOx1 catalysis. Using density functional theory calculation to examine the stereo-specific formation of the eight-membered ring endoperoxide, Chen et al. suggested that Tyr224 involvement might be essential to control the C_{21} position stereochemistry of the endoperxide chiral center (Ji & Chen, 2018).

In summary, although there was a significant progress in recent years on FtmOx1-catalyzed endoperoxidation reaction mechanistic studies, many of the proposed reaction intermediates in FtmOx1-catalysis (Fig. 10) remained to be characterized.

3. Fumitremorgin B oxidase (EAW25734)

In verruculogen biosynthesis, FtmE is responsible for the oxidative C-N bond formation. FtmE can accept both tryptostatin A (**3**) or tryptostatin B (**4**) as the substrate. Mechanistic details for this oxidative cyclization reaction still await to be explored. Recently, Ran et al. analyzed the genome of another verruculogen and fumitremorgin B producing strain, *N. fischeri* NRRL181 (Mundt et al., 2012). In *N. fischeri* NRRL181, it has the verruculogen biosynthetic gene cluster. The FtmOx1 homolog in *N. fischeri* NRRL181 ($FtmOx1_{nf}$) shares 95% sequence homology with *A. fumigates* FtmOx1 ($FtmOx1_{Af}$). Interestingly, sequence analysis of the *N. fischeri* genome identified another potential non-heme iron enzyme EAW25734 encoded by *NFIA_045530* gene in *N. fischeri* NRRL181 (Ran, Wohlgemuth, Xie, & Li, 2018). *NFIA_045530* gene is not located in the fumitremorgin gene cluster, and shares ~48% sequence identity with $FtmOx1_{Af}$ and $FtmOx1_{Nf}$. Biochemical characterization of recombinant EAW25734 protein using fumitremorgin B (**8**) as the substrate in the presence of α-ketoglutarate, ascorbic acid, and Fe^{2+}, did not lead to any product formation. These biochemical results suggest that EAW25734 protein from *N. fischeri* NRRL181 is not an endoperoxidase.

Ran et al. also examined other intermediates in the verruculogen biosynthetic pathway. Interestingly, EAW25734 could accept tryprostatin B (**4**, Fig. 12) as the substrate. In the presence of α-ketoglutarate, ascorbic acid, and Fe^{2+}, three products were observed in EAW25734 reaction (Fig. 12). These products were isolated and characterized by both NMR and mass spectrometry. For the major product, ^{1}H NMR information is most consistent with a migration of the double bond in the substrate at the C_{21}/C_{22} position to the C_{21}/C_{20} in the product (Fig. 12) and this conclusion was further confirmed by ^{13}C- and relevant 2D-NMR spectra. These spectroscopic information led to the assignment of 22-hydroxylisotryprostatin B (**41**) as the major product in EAW25734-catalysis, involving both a double bond migration and hydroxylation on the prenyl arm (Fig. 12). Whereas, the second product (**43**) is a minor product and accounts for only about 3.8% of the product mixture. NMR and mass spectrometry characterization of this compound led to the assignment of the structure with hydroxylation at the C_{14} position. The third product (**42**) accounts for ~27% of the product mixture. In this compound (**42**), the double bond migrated from the C_{21}/C_{22} position in the substrate to the C_{21}/C_{20} position in the product. At the same time, both C_{14} and C_{22} positions are hydroxylated. Both **41** and **43** could serve as the substrate for EAW25734 protein, while **41** is converted to **42** faster than that of **43**. According to this result, Ran et al. proposed that EAW25734 could catalyze multiple transformations (Fig. 12).

For the two above transformations, Ran et al. also examined the origin of oxygen in the hydroxyl groups. When the reaction was conducted under either $^{18}O_2$-enriched atmosphere or in ^{18}O-enriched water (97% purity for both $^{18}O_2$ and $H_2^{18}O$), very different results were observed for these two hydroxylation sites. First, when the reaction was conducted in $^{18}O_2$ atmosphere, there was 35% [^{18}O]-enrichment for the C_{22} hydroxylation (**41**, Fig. 12). On the contrary, there is nearly 95% [^{18}O]-enrichment in C_{14} position (**43**, Fig. 12). Consistent with the [$^{18}O_2$]-experimental results, when the reaction was conducted in [$H_2^{18}O$], incorporation of one

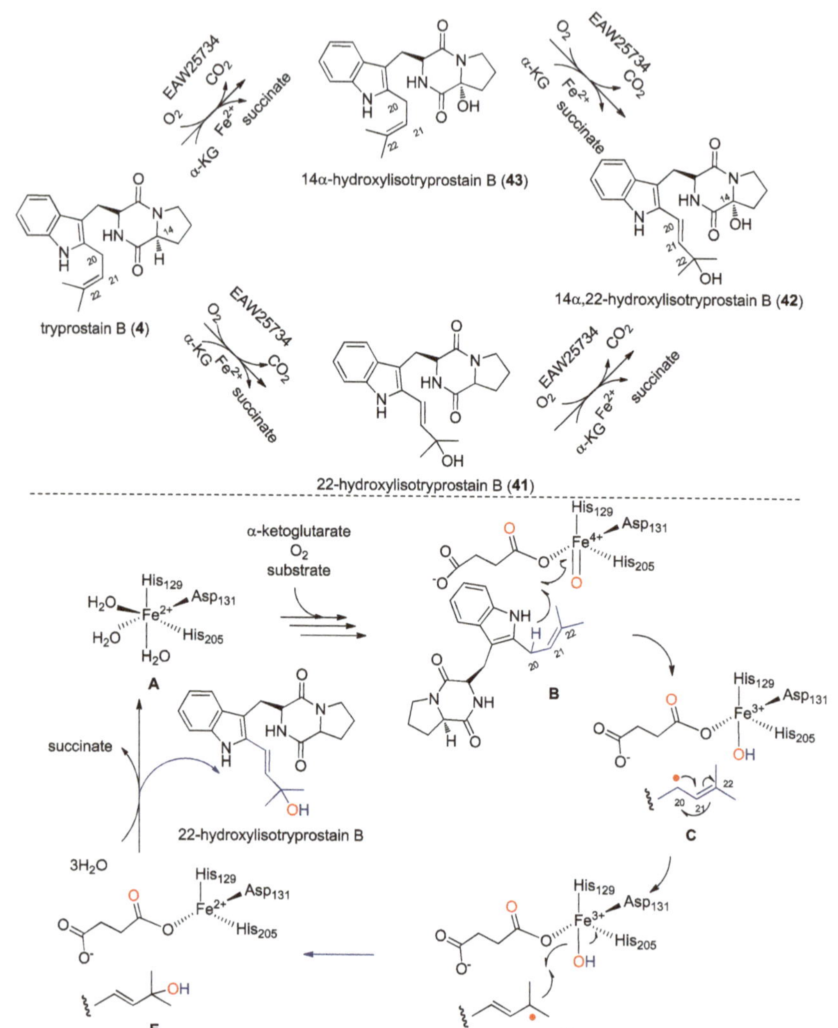

Fig. 12 Fumitremorgin B Oxidase (EAW25734). Top panel is the reaction and the bottom panel is the proposed mechanistic model.

[^{18}O]-atom into the hydroxyl group at the C_{22} position was observed, while no [^{18}O]-atom was incorporated at the C_{14} position. These are unusual results. If both C_{14} and C_{22} hydroxylations are coming from the same intermediate, similar levels of [^{18}O]-enrichments in both products were expected. Therefore, it is possible that these two reactions operate through two distinct mechanisms.

Thus far, there was no mechanistic study on these transformations yet. Ran et al. proposed a mechanism to explain the above results (bottom panel, Fig. 12). The α-ketoglutarate binding triggers O_2 activation to produce the Fe(IV)=O species, which abstracts a hydrogen at the C_{20} position of tryprostatin B (**4**) to form an allylic radical (**species E**, Fig. 12). The subsequent hydroxyl rebound to the C_{22} position results in the formation of the C_{22} hydroxylation product, while the Fe^{3+} center was reduced to Fe^{2+} (**species E**, Fig. 12). The formation of product **43** will be formed with a usual mechanism in which the hydrogen atom abstraction and hydroxyl rebound occurs at the same position. Notably, in the previous section of FtmE-catalyzed C-N bond formation reaction (Fig. 3), it was suggested that C-N bond is formed by a direct oxidative coupling reaction. However, feeding studies in *P. verruculosum* using (3*RS*)-[2-^{13}C] mevalonolactone as the source of the prenyl group indicated that there are significant levels of [^{13}C]-enrichment at both C_{28} and C_{29} positions (Fig. 8). This result suggests that from brevianamide F (**2**) to verruculogen (**1**) biosynthetic process (Fig. 2), one of the steps could lead to a significant level of bond rotation around the C_{26} and C_{27} positions (Vleggaar et al., 1993). There is currently no reported mechanistic study on the FtmE-catalysis. However, EAW25734-catalyzed transformation might be an alternative mechanism to be considered for FtmE-catalysis. The reaction goes through hydroxylation at the C_{22} position followed by a nucleophilic attack by nitrogen atom in the diketopiperazine ring and results in the formation of the C-N bond. Simultaneously, the double bond at C_{21}/C_{22} is regenerated by a dehydration reaction.

4. Novofumigatonin biosynthetic pathway

After the discovery of endoperoxidation catalyzed by FtmOx1 in verruculogen biosynthesis, Matsuda et al. discovered and reported another α-ketoglutarate dependent non-heme iron endoperoxidase, NvfI in novofumigatonin biosynthesis (Matsuda et al., 2018). Novofumigatonin (**59**, Fig. 13), isolated from *Aspergillus novofumigatus* IBT 16806 (CBS117520), is a fungal meroterpenoid with an unusual orthoester moiety (Rank et al., 2008). It was proposed that for novofumigatonin biosynthesis, it might involve a polyketide precursor 3,5-dimethylorsellinic acid (DMOA, **44**) (Ishikawa et al., 2016), which has been discovered to be a key intermediate for a diverse range of fungal meroterpenoids (Matsuda & Abe, 2016). The characterization of novofumigatonin biosynthetic pathway reveals that it

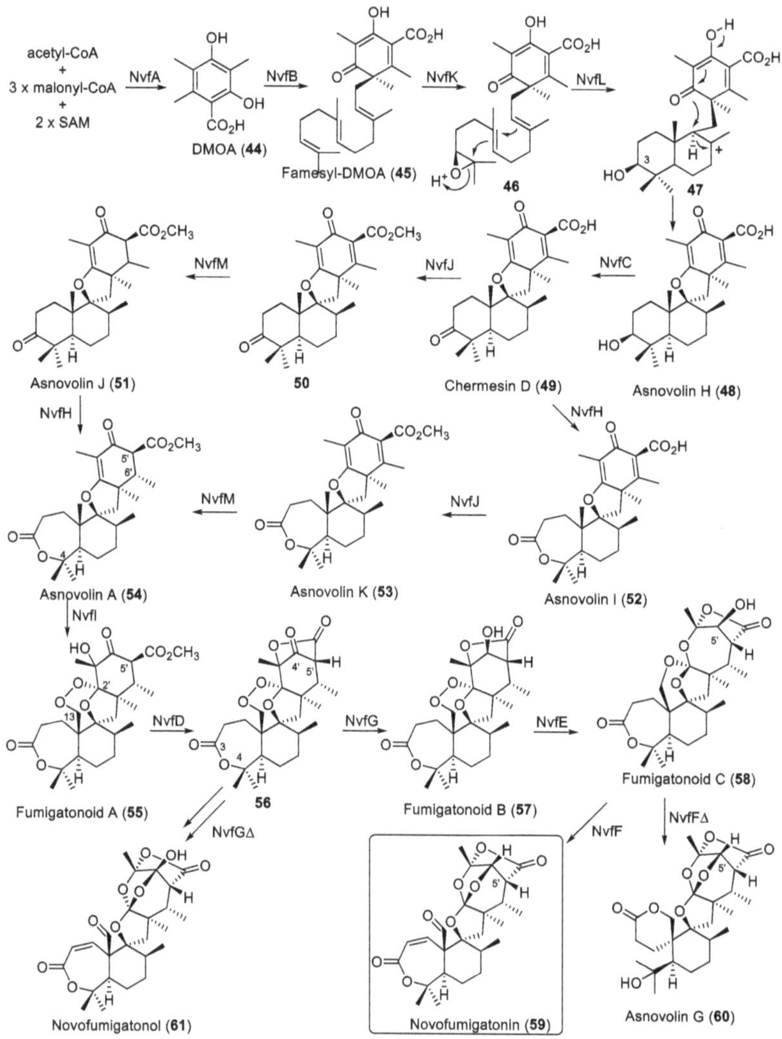

Fig. 13 Novofumigatonin biosynthetic pathway.

involves an endoperoxide containing intermediate (**55**), and its biosynthesis is catalyzed by an α-ketoglutarate-dependent non-heme iron enzyme NvfI (Fig. 13). Based on the assumption that novofumigatonin biosynthesis involves DMOA (**44**, Fig. 13) as an intermediate, they analyzed the *A. novofumigatus* genome and identified a potential biosynthetic gene cluster. When the DMOA synthase gene (*NvfA*) was deleted, it completely abolishes novofumigatonin biosynthesis, providing the initial support for the assignment of the novofumigatonin biosynthetic gene cluster. After this cluster was transferred to a heterologous host, *Aspergillus oryzae*, successful production of novofumigatonin provided additional evidence to support this assignment.

After the DMOA (**44**) is synthesized by a polyketide synthase (NvfA), a farnesyl group is then transferred to the DMOA core by the NvfB protein. The NvfB functional assignment was confirmed by both characterization of the *nvfB* deletion mutant and the results from co-expression of *nvfB* along with a known DMOA-synthase gene *andM* (Matsuda, Wakimoto, Mori, Awakawa, & Abe, 2014) in the *A. oryzae* NSAR1 strain (Jin, Maruyama, Juvvadi, Arioka, & Kitamoto, 2004). Both *nvfK*Δ and *nvfL*Δ mutants accumulated compounds **44** as the key metabolite and NvfK was assigned as the epoxidase, which converts farnesyl-DMOA (**4**) to an epoxide **46**. NvfL is the cyclases, which catalyzes the cyclization by protonating the epoxide **46** to open the ring to create a cation intermediate to initiate the cyclization cascade to produce asnovolin H (**48**) as the product.

To identify the genes in the mid-stage of the novofumigatonin biosynthetic pathway, three more mutants were characterized: *nvfC*Δ, *nvfH*Δ, and *nvfM*Δ strains. Mutation of these three genes abolished the production of both novofumigatonin (**59**) and asnovolin A (**54**), suggesting that these genes are involved in the biosynthesis. The *nvfH*Δ strain accumulated asnovolin J (**51**). NvfH is a flavin monooxygenase, which is responsible for the insertion of an oxygen atom between the C-3 and C-4 positions, a Baeyer–Villiger monooxygenase (BVMO) reaction to converted compound **51** to **54**. In the *nvfM*Δ strain, asnovolin K (**53**) was accumulated. In compound **53**, relative to asnovolin A (**54**), there exists a double bond between C-5′ and C-6′, which indicates that NvfM is an enoylreductase, catalyzing the reduction of the double bond between C-5′ and C-6′. The assignment of function of NvfC was supported by heterologous expression of *andM*, *nvfB*, *nvfK*, *nvfL*, and *nvfC* genes in the *A. oryzae* heterologous expression system. In the absence of *nvfC* gene, compound **48** was produced, while when *nvfC* gene was included, both compound **48** and chermesin D (**49**) were accumulated. With this information, NvfC is assigned as a dehydrogenase, catalyzing the oxidation of C_3-OH in asnovolin H (**48**) to chermesin D (**49**). By integrating these genetic results, two potential pathways were envisioned for the steps responsible for converting chermesin D (**49**) to asnovolin A (**54**). In both potential pathways, methylation (NvfJ-catalysis) occurs before the enol reduction step (NvfM-catalysis).

The late-stage biosynthetic enzymes for novofumigatonin (**59**) biosynthesis have also been assigned based on genetic studies. In *nvfD*Δ, *nvfE*Δ, *nvfF*Δ, *nvfG*Δ, and *nvfI*Δ mutants, there is no novofumigatonin **59** production, while asnovolin A (**54**) accumulates, which suggests that these five genes are involved in converting asnovolin A (**54**) to novofumigatonin (**59**). In *nvfD* and *nvfE* mutants, compounds **55** and **57** were accumulated, respectively. Compound **55** was also produced by the *nvfD*Δ, *nvfE*Δ, *nvfF*Δ, *nvfG*Δ quadruple-mutant strain, implying that compound **55** is a product of NvfI. In addition, interpretation of the NMR spectra suggested that both compounds fumigatonoids A (**55**) and fumigatonoids B (**57**) contained an endoperoxide linkage between the C_{13} and $C_{2'}$ carbon. Therefore, NvfI is an endoperoxidase and sequence analysis suggests that NvfI is α-ketoglutarate dependent non-heme iron enzyme. This is the second reported α-ketoglutarate dependent non-heme iron enzyme endoperoxidase. For metabolites isolated from the *nvfE*Δ, *nvfF*Δ, and *nvfG*Δ strains, the methyl ester group present in fumigatonoids A (**55**) is missing. These results led to the assignment of NvfD as the enzyme responsible for the lactonization reaction (**55** → **56** transformation). Sequence analysis suggests that NvfD is an α/β hydrolase. There are literature examples of having an α/β hydrolase with an epimerase activity (Gaudelli & Townsend, 2014). NvfD therefore was assigned as the epimerase that converting **55** to its $C_{5'}$ epimer, which then undergoes spontaneous or NvfD-catalyzed lactonization to produce compound **56**. Compound **56** is then reduced by a ketoreductase NvfG to produce **57**. For the next two enzymes (NvfE and NvfF), sequence analysis suggests that both of them are α-ketoglutarate dependent non-heme iron enzymes. In the *nvfF*Δ strain, compound **60** was isolated. Compound **60** is most likely the product isomerized from a real biosynthetic intermediate, fumigatonoid C (**58**), which will be discussed in the next section.

The assignment of NvfI as an endoperoxidase was also supported by biochemical evidences. Using recombinant NvfI protein, in the presence of α-ketoglutarate and O_2, asnovolin A (**54**) was converted to fumigatonoid A (**55**). Notably, the presence of excess Fe^{2+} around renders fumigatonoid A **55** stable resulting in the decomposition of the endoperoxide. In NvfI catalysis, there is also a by-product, which is the monohydroxylated product asnovolin B. To study the transformation of endoperoxide to orthoester transformation, the other two proposed non-heme iron enzymes in novofumigatonin **59** biosynthetic gene cluster have also been characterized. In NvfE enzymatic reaction, when compound **57** was used as the substrate, two products were isolated. The first one is compound **60**, which has been isolated from *nvfF*Δ mutant strain. The other product is compound **58**. Relative to compounds **59** and **60**, compound **58** has a different orthoester moiety, in which the hydroxyl group at $C_{5'}$ position is not involved in the orthoester formation. In the presence of formic acid, compound **58** was transformed into compound **60** easily. These results led Matsuda et al. to propose that compound **58** is the real biosynthetic intermediate, while compound **60** is the isomerized product from compound **58**. This assignment was supported by the biochemical properties of the next α-ketoglutarate-dependent non-heme iron enzyme, NvfF. In the presence of Fe^{2+} and α-ketoglutarate, NvfF can efficiently convert compound **58** to novofumigatonin **59**, while it does not take compound **60** as the substrate.

Therefore, compound **58** was assigned as the native novofumigatonin **59** biosynthetic intermediate.

Among these three predicted α-ketoglutarate dependent non-heme iron enzymes, NvfE is also standout. It catalyzes an isomerization reaction. However, the activity does not dependent on the presence of α-ketoglutarate or Fe^{2+}, which led Matsuda et al. to propose that NvfE is a cofactor/cosubstrate-free enzyme. Mutation of the three potential Fe^{2+} ligands (H152, D154, and H234) led to a reduced level of activity, while the activity was not completely abolished. Therefore, all of these three non-heme iron enzymes in novofumigatonin biosynthetic pathway are unique.

Matsuda et al. have also proposed a mechanism to account for the endoperoxidation reaction observed for NvfI-catalysis. An Fe(IV)-oxo species is generated by following the typical α-ketoglutarate mediated O_2 activation mechanism (Fig. 5A). This reactive species then abstracts a hydrogen atom at the C_{13} position to generate radical species **62** (Fig. 14). This carbon-centered radical reacts with a molecular oxygen (O_2) to produce the peroxy radical **63**. The peroxy radical **63** then reacts with the alkene at its $C_{2'}$ position to create endoperoxide functional group in intermediate **64**. Different from FtmOx1-catalysis, hydroxyl-rebound at the $C_{3'}$ position completes the reaction cycle to provide **55** as the final product (Fig. 14). This proposed NvfI endoperoxidation mechanism is distinct from FtmOx1-catalysis (Fig. 10) (Kato et al., 2011; Steffan, Afiyatullov, et al., 2009; Yan et al., 2015). In FtmOx1-catalysis, C_{26} radical is quenched either by a reductant (e.g. ascorbate) or by going through the second reaction, the oxidation of C_{13}-hydroxyl group to a keto (Fig. 10). In NvfI-catalysis, the reaction is quenched by the hydroxyl rebound mechanism (Fig. 14).

It is well known that Fe^{2+} causes the cleavage and rearrangement of endoperoxide linkages in a variety of compounds such as the antimalarial agent artemisinin (Yaremenko, Vil, Demchuk, & Terent'ev, 2016). Matsuda

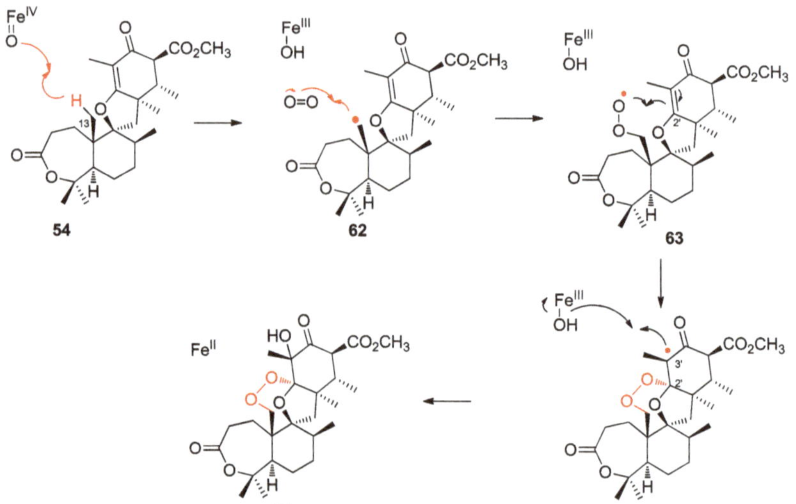

Fig. 14 Proposed endoperoxidation mechanism in NvfI-catalysis.

et al. proposed that the NvfE-catalyzed orthoesterification is also initiated by the Fe^{2+}-mediated cleavage of the endoperoxide (Fig. 15A). First, NvfE-active Fe^{2+} center initiates the endoperoxide cleavage to produce an Fe^{3+} species and an oxygen-centered radical **65**, which undergoes β-scission to a carbon-centered radical **66**. Radical **66** is oxidized by Fe^{3+} to form a carbocationic species **67**. The carbocation species **67** is quenched by two rounds of heterocyclization to afford the orthoester **58**. The details for the NvfF reaction are still obscure at this point. NvfF-catalysis involves two catalytic cycles, dehydrogenation reaction to introduce a double bond between the C_1 and the C_2 positions, and the re-arrangement of orthoesters and the formation of an aldehyde at the C_{13} position (Fig. 15B). Homologs of NvfF are found in many DMOA-derived meroterpenoid biosynthetic pathways, and most of them catalyze multi-step reactions (Matsuda et al., 2014, 2016; Wu, Meng, & Tang, 2016). The proposed reaction mechanism for NvfE-catalysis is outlined in Fig. 15B. In this model, it was suggested that Fe(IV)=O species generated from α-ketoglutarate mediated O_2 activation abstracts a hydrogen atom from the C_{13} position of compound **58** to mediate hydroxylation at this position, or mediate the dehydrogenation between the C_1 and C_2 positions. The resulting hemiacetal **68** is spontaneously converted to compound **69** with both an aldehyde and hemiorthoester functional groups. The transformation from hemiorthoester **70** to the orthoester in novofumigatonin **59** was proposed to go through a cationic

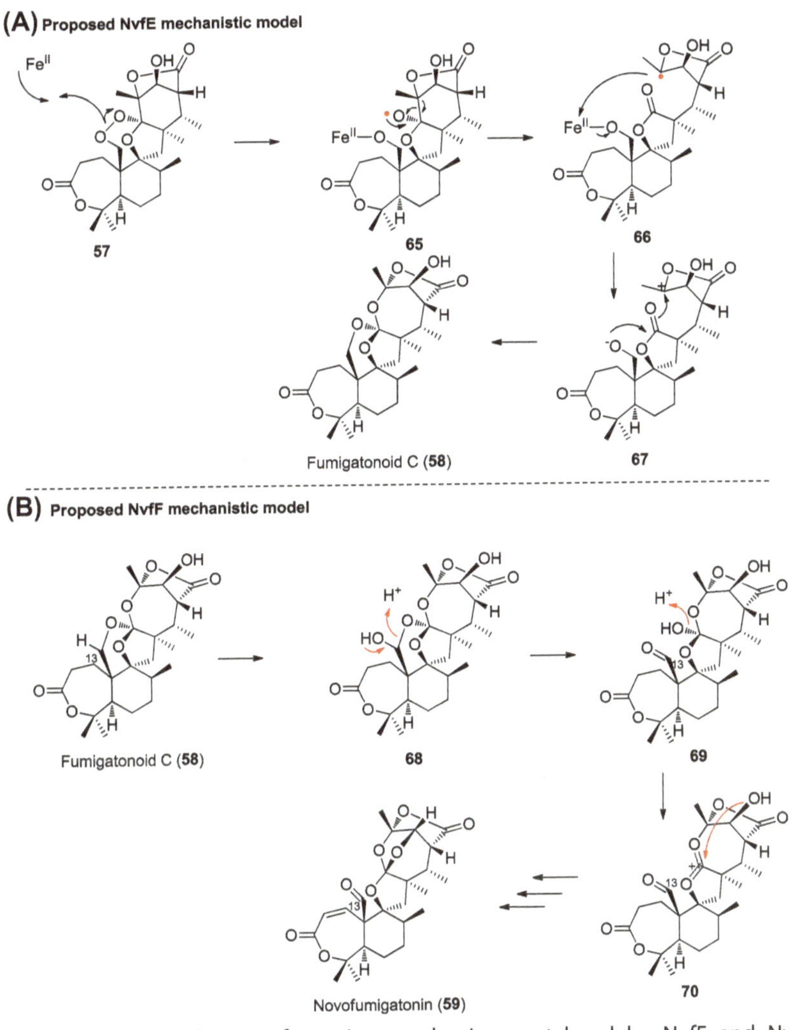

Fig. 15 Proposed orthoester formation mechanisms catalyzed by NvfE and NvfF-catalysis.

intermediate. All of these are novel and complicated transformations and mechanistic details still await to be explored.

5. EasH: an oxidase involved in cycloclavine biosynthesis

Cycloclavine (**74**) belongs to ergot alkaloids (Fig. 16), a class of fungal-derived natural products with a wide range of biological activities and

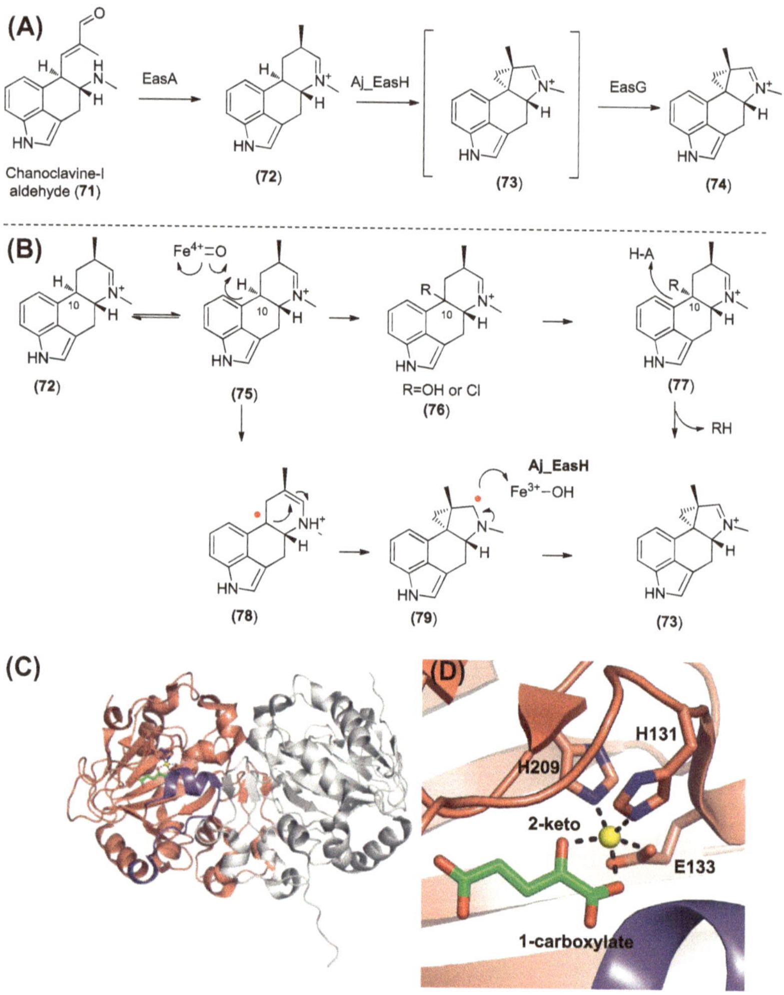

Fig. 16 Proposed function of EasH, an oxidase involved in cycloclavine biosynthesis. (A) reaction pathway; (B) Proposed EasH reaction mechanism; C & (D) Crystal structural information of EasH.

pharmaceutical applications (Jakubczyk, Cheng, & O'Connor, 2014; Wallwey & Li, 2011). Cycloclavine has a unique fused 5-3 ring system. Recently, it was discovered that a non-heme iron enzyme EasH from *Aspergillus japonicus* (Aj_EasH) is responsible for the oxidative C-C bond formation of this 5−3 ring system using a 6-membered ring in compound **72** as the starting material (Furuta, Koike, & Abe, 1982; Jakubczyk et al., 2015). In conjunction with a flavoenzyme (EasA) and an NADPH-dependent oxidoreductase (EasG), these three enzymes (EasA, Aj-EasH, and EasG) convert chanoclavin-I aldehyde (**71**) to cycloclavine (**74**) (Cheng et al., 2010; Jakubczyk et al., 2015; Matuschek, Wallwey, Xie, & Li, 2011). Recently, the crystal structure of Aj_EasH · α-ketoglutarate binary complex has been reported (Jakubczyk, Caputi, Stevenson, Lawson, & O'Connor, 2016). The Aj_EasH iron-center is coordinated by H131, E133, H209 and α-ketoglutarate binds to the iron-center bidendately (Fig. 16D). A short helical segment (residues 52–72) as part of a surface loop functions as a "lid" to close the Aj_EasH active site. Thus far, structure of the Aj_EasH•substrate or the Aj_EasH · product binary complex was not yet available. Due to the hydrophobic nature of the substrate, the likely substrate binding pocket of Aj_EasH is easy to predict and Jakubczyk et al. proposed that a spacious and hydrophobic site (the pocket formed by Tyr65, Tyr69, Val236, Thr115, Leu119, Leu275, Leu70, Ile227 and Gly136) is most likely the substrate binding pocket.

In the EasA- Aj_EasH-EasG cascade reaction, chanoclavine-I aldehyde **71** is the initial substrate of the reaction cascade. It is oxidized by the flavoenzyme EasA to form an iminium species **72**. Jakubczyk et al. proposed that compound **73** (the enamine tautomer of **72**), is the actual substrate for Aj_EasH. Using Autodock program (Morris et al., 1998), followed by manual inspection of the substrate binding in the proposed substrate binding pocket, Jakubczyk et al. created a docking model of Aj_EasH · substrate binary complex. Two potential mechanisms have been proposed for this unprecedented transformation (Fig. 16B). In the first mechanism (Jakubczyk et al., 2015), the Fe(IV)=O species abstracts a hydrogen atom from the C_{10} position to form a C_{10} based radcial, which then followed by hydroxylation or halogenation at the C_{10} position (**76**, Fig. 16). In the subsequent step (the rearrangement of the carbon skeleton to form the 5-3 ring system), the C_{10} hydroxyl or halide moiety functions as a leaving group (**76** → **77**, Fig. 16B). In the alternative mechanism, the formation of the C_{10}-based substrate radical is followed by a radical-based skeleton re-arrangement to produce product based radical (**79**, Fig. 16B), which is then oxidized to afford the iminium product **73** (Fig. 16B). Cyclopropyl group biosynthesis has been reported to occur *via* either carbocationic or carboanionic intermediates in the terpenoid biosynthesis (Thibodeaux, Chang, & Liu, 2012). The Aj_EasH catalysis represents an alternative mechanism for the production of this common functional group in many natural products.

6. Orthoester formation in hygromycin B biosynthesis

Hygromycin B (**81**) is an aminoglycoside antibiotic produced by *Streptomyces hygroscopicus*. Hygromycin B perturbs protein synthesis (Carter et al., 2000) by disrupting the bacterial 30S ribosomal subunit (Pfister, Risch, Brodersen, & Bottger, 2003; McGaha & Champney, 2007) and eukaryotic ribosome activities (Eustice & Wilhelm, 1984). Hygromycin B belongs to a class of natural products which contain a unique orthoester moiety (Fig. 17) and this is a class of natural products that are rich in biological activities (Chau et al., 2011, 2013; Jal & Khora, 2015). The spirocyclic ortho-δ-lactone functionality in hygromycin B has also been identified in other oligosaccharide antibiotics, including avilamycin (**82**) (Hofmann et al., 2005; Weitnauer et al., 2001, 2002), and everninomicin (**83**) (Arenz et al., 2016; Vey et al., 2010). Thus far, there are no mechanistic studies on the biosynthesis of orthoesters in this class of natural products (Tang, Zou, Watanabe, Walsh, & Tang, 2017). In addition, due to the complexity of the system, orthoester formation reactions in these oligosaccharide antibiotics have not yet been defined, except a recent report for hygromycin B (**81**) biosynthesis (Li et al., 2018). It was proposed that the orthoester formation was catalyzed by α-ketoglutarate dependent non-heme iron enzymes (McCulloch et al., 2015).

Recently, Li et al. provided evidences supporting the role of an α-ketoglutarate dependent non-heme iron enzyme (HygX) in the construction of the orthoester in hygromycin B biosynthesis (Li et al., 2018). Based on the predicted hygromycin B biosynthetic gene cluster (GenBank accession number AJ628642.1) (McCulloch et al., 2015), Li et al. generated the *hygX* deletion strain from a hygromycin B-producing strain *S. hygroscopicus* subsp. *hygroscopicus* DSM 40578 (Komaki et al., 2017). Metabolites from Δ*hygX* were analyzed by liquid chromatography coupled with electrospray ionization high-resolution mass spectrometry (LC–ESI-HRMS). Indeed, in the *hygX* deletion strain, the production of hygromycin B (**81**) was abolished, which provides initial evidence supporting the involvement of HygX in hygromycin B (**81**) biosynthesis. In their subsequent experiments, they add a copy of *hygX* gene under its native promoter to the *hygX* deletion strain and it led to the successful production of hygromycin B (**81**). A combination of these two sets of experiments provided evidences supporting the involvement HygX gene in hydromycin B biosynthesis. Carefully analysis of the metabolites in Δ*hygX* strain led to the

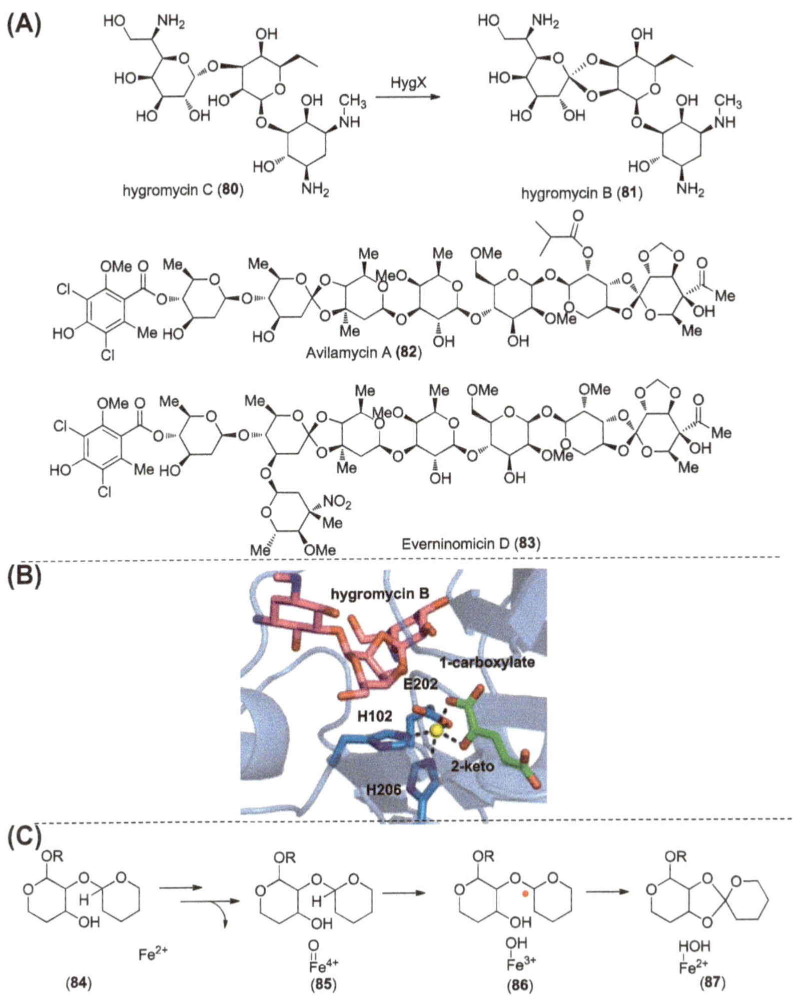

Fig. 17 Orthoester formation catalyzed by non-heme iron enzymes. (A) a few examples of orthoester containing natural products; (B) Crystal structure of HygX in hygromycin B biosynthesis; (C) A proposed mechanism for HygX-catalyzed orthoester formation reaction.

identification of a new metabolite, which was named as hygromycin C (**80**, Fig. 17). Hygromycin C (**80**) was also detected as a minor component in extracts of the wild-type strain. To test whether hygromycin C (**80**) is the substrate for HgyX to construct the orthosoester functionality, recombinant HgyX was incubated with hygromycin C (**80**) in the presence of

Fe^{2+} and α-ketoglutarate. Indeed, hygromycin C was successfully converted into hygromycin B (**81**), which provided direct evidence supporting that HygX is responsible for the construction of the orthoester functionality (Fig. 17A) (Li et al., 2018).

The crystal structure of HygX along with the potential orthoester formation enzymes in avilamycin (**82**), and everninomicin (**83**) biosyntheses have also been reported recently (McCulloch et al., 2015). To identify the genes responsible for the orthoester construction in avilamycin and everninomicin biosyntheses, McCulloch et al. first analyzed five orthosomycin biosynthetic gene clusters available in GenBank (McCulloch et al., 2015): *ava* (avilamycin biosynthesis from *Streptomyces mobaraensis*), *avi* (avilamycin biosynthesis from *Streptomyces viridochromogenes Tü57*), *evd* (everninomicin biosynthesis from *Micromonospora carbonacea* var. *aurantiaca*), *eve* (everninomicin biosynthesis from *M. carbonacea* var. *africana*), and *hyg* (hygromycin B biosynthesis from *S. hygroscopicus*). The *avi* and *hyg* gene clusters have functional annotations already. Because the orthoester formation was proposed to be catalyzed by α-ketoglutarate dependent non-heme iron enzymes, they have analyzed all of the 13 predicated α-ketoglutarate dependent non-heme iron enzymes in these clusters. They then conducted phylogenetic analysis of these 13 genes and obtained the crystal structures of the one most likely responsible for the orthoester formation (AviO1, EvdO2 EvdO1, and HygX). Because the functions of the other enzymes have not yet been confirmed, we only summarized the HygX structural information here.

When the HygX structure was reported, its biochemical function has not yet been validated by biochemical studies (Li et al., 2018). Due to the lack of biochemical results, McCullonch et al. co-crystallized HygX with its proposed product, hygromycin B (**81**) and reported the structure of the tertiary complex (HygX·α-ketoglutarate·hygromycin B) at 1.6 Å resolution (Fig. 17B). In this structure, one of the bridging oxygen of hygromycin B (**81**) is pointing toward the metal center (Fig. 17B). Comparison of this structure with that of clavaminate synthase (Zhang et al., 2000) aligns an anomeric carbon of hygromycin B (**81**) with the carbon atom of proclavaminate that undergoes hydrogen atom abstraction. Therefore, McCulloch et al. proposed that hygromycin C (**80**) anomeric carbon is most likely the site for the hydrogen atom abstraction, followed by hydroxyl rebound to form the orthoester (Fig. 17C), as in the case of fosfomycin epoxidase (HppE) (Chang et al., 2013; Higgins, Yan, Liu, Liu, & Drennan, 2005; Liu, Ho et al., 2001).

7. Nogalamycin biosynthesis

Nogalamycin (**88**, Fig. 18) is a member of the anthracycline family of antibiotics. Its biosynthesis attracts particular interests in recent years because it has an unusual carbohydrate unit, nogalamine, which is attached both *via* a carbon—carbon bond and a classical O-glycosidic linkage to the aglycone

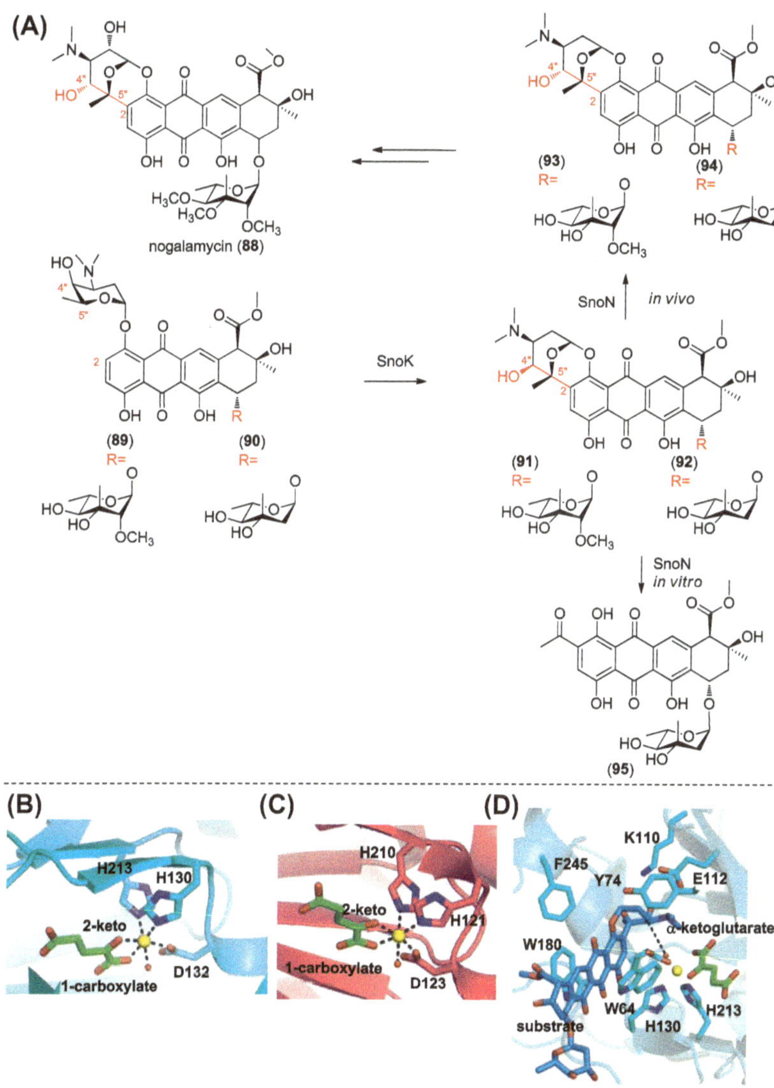

Fig. 18 Non-heme iron enzymes in nogalamycin biosynthetic pathway. (A) Part of the biosynthetic pathway; (B)—(D) SnoK and SnoN structural information.

(colored in red in **88** of Fig. 18). The formation of C5′′-C2 bond is a mechanistically intriguing question (Siitonen et al., 2016). In addition, results from recent genetic studies indicated that rhodosamine is actually the precursor to nogalamin biosynthesis (Siitonen, Blauenburg, Kallio, Mantsala, & Metsa-Ketela, 2012). The stereochemistries at C4′′-position are different in rhodosamine and nogalamin, which implies that an eperimization process is involved (C_4'' stereochemistries in **89**–**92** relative to **93** and **94**, Fig. 18). Based on functional characterizations of most of the nogalamycin biosynthetic gene clusters (Siitonen, Claesson et al., 2012, Siitonen, Blauenburg et al., 2012), recent efforts were focused on two α-ketoglutarate dependent non-heme iron enzymes (SnoN and SnoK) responsible for these two transformations (Siitonen et al., 2016). Coexpression of *snoN*, along with a cosmid containing the early steps of the nogalamycin (pSnogaori) in *S. albus*, resulted in the production of two anthracyclines, **93** and **94**. Structural analysis by NMR and mass spectrometry indicated that compound **93** was identical to **91** except for the stereoinversion at the C_4'' position, which provides that first hint that *snoN* encodes a C_4'' epimerase. In cosmid pSnogaori, when *snoK* gene was deleted, compounds **89** and **90** were accumulated. Structural analysis indicated that in these two compounds (**89** and **90**), the $C_5''-C_2$ bond was not present. Therefore, SnoK is responsible for this oxidative C-C bond formation (Fig. 18).

These assignments were confirmed by *in vitro* biochemical characterizations. In the presence of Fe^{2+}, α-ketoglutarate, and ascorbic acid, SnoK catalyzes the $C_5''-C_2$ cross-linking in compound **89** to form compound **91**. Quantitative analysis indicated that the stoichiometry for the amount of α-ketoglutarate consumed and the amount of compound **91** formed is approximately 1:1 in ratio. Under similar conditions, when compound **92** was used as the substrate, SnoN-reaction produced several products and one of the compounds identified is compound **95**, which loses most of the amino sugar portion and only contained a ketone functional group at the C_2 position.

In SnoN biochemical analysis, it was found that the stoichiometry for the amount of α-ketoglutarate consumed relative to the amount of substrate **92** was 1:2, which is unexpected.

Siitonen et al. have also determined the crystal structures of SnoN and SnoK in complex with α-ketoglutarate and mononuclear iron (Fig. 18). Similar to other α-ketoglutarate-dependent non-heme iron enzymes, the central core of both SnoN and SnoK is a jelly-roll or β-sandwich fold (Gao et al., 2018). The mononuclear iron center is SnoN are coordinated by His130, Asp132, and His213 (Fig. 18B), while the SnoK iron center is coordinated by His121, Asp123, and His210 (Fig. 18C). In this structure, the α-ketoglutarate coordinates to the metallo-center bidentately to form the octahedral geometry. They have also obtained the structures of SnoN in complex with α-ketoglutarate and substrate **92** (Fig. 18D). The aromatic anthracycline ring system is sandwiched between Trp64 and Trp180 and surrounded by Tyr74 and Phe245. The rhodosamine moiety of **92** is located close to the iron center. The dimethylamino group is within hydrogen bond distance to Glu112, and the C_4'' hydroxyl group is hydrogen bonded to Lys110. These interactions place the substrate in a position where the H_4'' hydrogen atom points toward the iron center for reaction. Thus far, structure of SnoK in complex with its substrate was not yet available.

To explain these two novel transformations, Siitonen et al. have proposed a tentative mechanism (Fig. 19). Like other oxygenases, SnoK undergoes oxidative decarboxylation of α-ketoglutarate to form the Fe(IV)=O species (**species 96** and **98** in Fig. 19). In SnoK-catalysis, two hydrogen

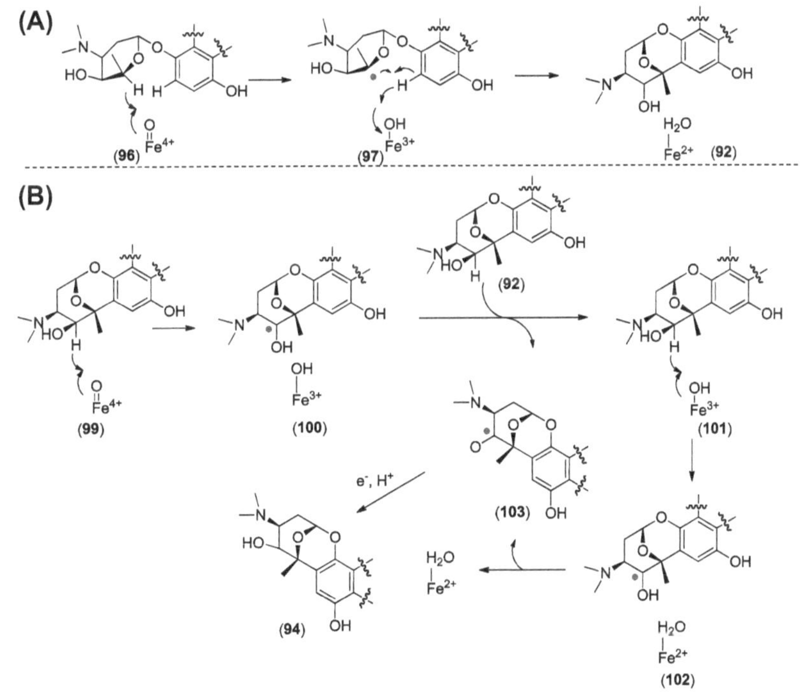

Fig. 19 Proposed mechanistic models for SnoK and SnoN-catalysis in nogalamycin biosynthesis. (A) Proposed SnoK mechanism; (B) Proposed SnoN mechanism.

atoms at the $C_{5''}$ and C_2 positions are removed sequentially to form the C-C bond in compound **91** or **92**. Siitonen et al. proposed that Fe(IV) = O abstracts the $C_{5''}$ hydrogen atom from the substrate to produce a substrate radical (**97**, Fig. 19A). The intramolecular ring closure is then accomplished after attack of the aromatic C_2 on the radical center to form the new C—C bond in the product (**97** → **92**, Fig. 19A). This is a chemically challenging transformation. The authors also compared this mechanistic model with P450 catalyzed C-C bond formation, in which phenoxy radicals are involved. The mechanism of C—C bond formation by SnoK is mechanistically different from intramolecular phenol radical coupling reactions catalyzed by P450 enzymes (Grobe et al., 2009; Ikezawa, Iwasa, & Sato, 2008). Further characterization will be needed to understand this novel transformation.

In SnoN-catalysis, there is no net redox-state change in the substrate as this is an epimerization reaction. In carbapenem biosynthesis, there is an α-ketoglutarate dependent non-heme iron enzyme CarC, which is a bifunctional enzyme catalyzing both epimerization and a subsequent desaturation reaction (Chang et al., 2014). In CarC-catalysis, the Fe(IV)=O species abstracts a hydrogen atom from the substrate to form the substrate-based radical, which is then quenched by Tyr165. External reductants are essential to CarC-catalysis. In SnoN, the active site cavity of SnoN contains three potential radical formation residues (Trp64, Tyr74, and Trp180). However, none of these residues are structurally equivalent to Tyr165 in CarC. Mutagenesis of any of these residues to phenylalanines did not significantly reduce SnoN activity. Thus far, the mechanistic details for SnoN-catalysis remain elusive. However, reported results suggested that SnoN clearly undergoes a mechanism different from CarC-catalysis. Based on the ratio between α-ketoglutarate and substrate, the authors proposed a tentative model in which the two substrate-based radicals are quenched by external reductants (Fig. 19B). In this mechanistic model, the product based radical **103** was released to solution and quenched by a reductant. In each SnoN catalytic cycle, it produces two molecules of radical **103**, which explains the 1:2 ratio between α-ketoglutarate consumed and the product **94** formed in this reaction.

8. Ergothioneine and ovothiol biosynthesis

Ergothioneine (**109**) and glutathione are among the most abundant small molecular thiols in animals. Ergothioneine's reduction potential

($E_0{'} = -0.06$ V (Hartman, 1990; Hand & Honek, 2005; Fahey, 2001)) is significantly higher than that of glutathione ($E_0{'} = -240$ mV (Scott, Duncan, & Ekstrand, 1963)). A combination of ergothioneine and glutathione offers cellular protection over a wide-range of oxidative stress conditions by removing reactive oxygen species (ROS) and reactive nitrogen species (RNS) and their functions are complementary in nature (Cheah & Halliwell, 2012; Aruoma, Whiteman, England, & Halliwell, 1997; Franzoni et al., 2006; Rougee, Bensasson, Land, & Pariente, 1988; Murphy et al., 2011). **First**, they react with ROS and RNS species with their preferences. **Second**, their anti-oxidation reactions are different (D'Onofrio et al., 2016; Song et al., 2015). The end oxidation product of ergothioneine is hercynine (**105**, Fig. 20A) and involves ergothioneine sulfenic and sulfinic acid as oxidation intermediates. In this process, sulfur oxidation states change from a "−2" to a "+4" state. In contrast, glutathione is normally oxidized to its disulfide form. Therefore, under oxidative stress conditions, and at the same concentration, ergothioneine is more efficient in eliminating oxidative species. Due to this superior anti-oxidative property, ergothioneine is applied as a protective reagent in organ-transplant (Arduini, Eddy, & Hochstein, 1990; Bedirli et al., 2004; Cheah & Halliwell, 2012; Fahey, 2001; Hand & Honek, 2005; Sakrak et al., 2008).

In humans, ergothioneine accumulates up to 2 mM in many parts of our body, with the highest concentration in erythrocytes, bone marrow, liver, kidney, seminal fluid, and the lens and cornea of the eyes (Leone & Mann, 1951; Melville, Horner, & Lubschez, 1954; Shires, Brummel, Pulido, & Stegink, 1997). Such accumulation is achieved by an ergothioneine-specific transporter (OCTN1) (Grigat et al., 2007; Grundemann et al., 2005; Kato et al., 2010; Nakamura, Yoshida, Yabuuchi, Maeda, & Tamai, 2008), and the concentrations of ergothioneine in various tissues correlate with the abundance of *OCTN1* mRNA in these cells (Grigat et al., 2007; Kato et al., 2010; Nakamura et al., 2008; Paul & Snyder, 2010; Taubert, Jung, Goeser, & Schomig, 2009). Besides functioning as an anti-oxidant, ergothioneine may also offer protection against radiation, modulating immune responses, heavy metal detoxification (e.g., Cu^{2+} detoxification), and regulating mitochondrial functions (Cheah & Halliwell, 2012). Ergothioneine has also been suggested to be beneficial to rheumatoid arthritis (Taubert et al., 2006; Tokuhiro et al., 2003), Crohn's disease (Leung et al., 2006; Peltekova et al., 2004), and diabetes (Bastard et al., 2002; Ridker, Rifai, Stampfer, & Hennekens, 2000). Ergothioneine can pass through the blood-brain barrier and protect against neurodegenerative

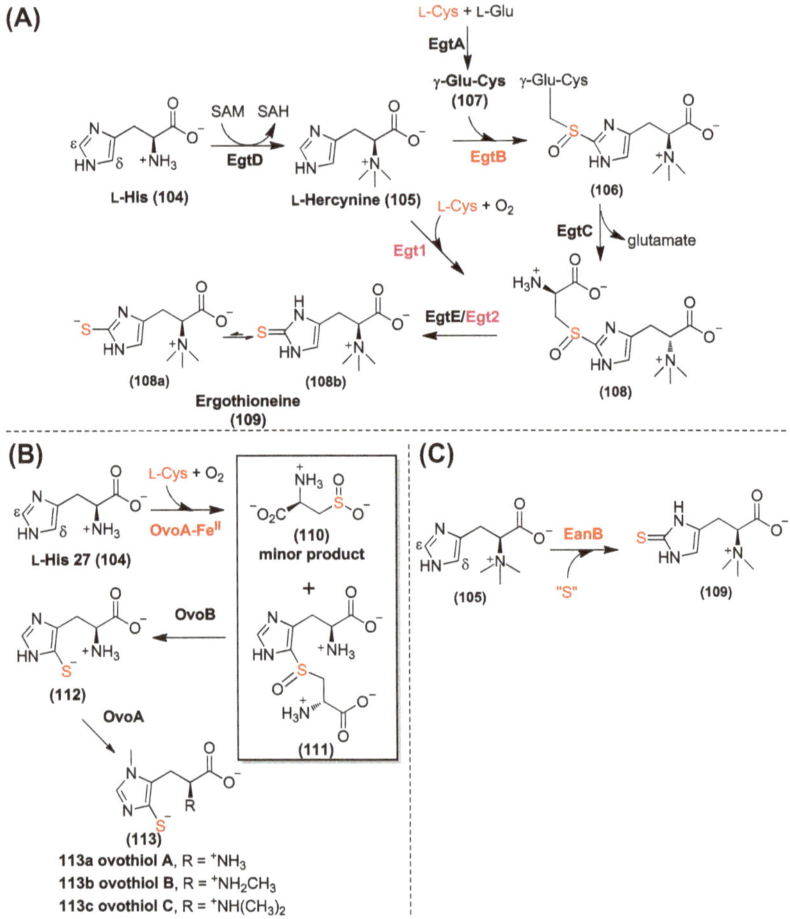

Fig. 20 Ergothioneine and ovothiol biosynthetic pathways. (A) Two aerobic ergothioneine biosynthetic pathways; (B) Ovothiol biosynthetic pathway; (C) Anaerobic ergothioneine biosynthetic pathway.

diseases (Briggs, 1972; Crossland, Mitchell, & Woodruff, 1966; Kaneko et al., 1980; Moncaster, Walsh, Gentleman, Jen, & Aruoma, 2002; Nakamura et al., 2008). Ergothioneine treatment inhibits the binding of monocytes to the endothelium by decreasing the expression of adhesion molecules (Martin, 2010), which might explain ergothioneine's protective effects to cardiovascular disorders (Libby, Ridker, & Hansson, 2011). For these reasons, there is an increasing interest of studying its biosynthetic details and developing methods to produce ergothioneine through fermentation.

Recently, we have summarized the biosynthesis of ergothioneine and ovothiol, which involves unprecedented sulfur-transfer strategies (Naowarojna, Cheng, et al., 2018). Three different ergothioneine biosynehtic pathways have been elucidated (Fig. 20A and C). In *mycobacterium* ergothioneine biosynthesis, it contains a five-gene cluster, designated *egtABCDE* (EgtA – EgtE genes). EgtB is the most critical enzyme in the mycobacterium biosynthetic pathway. It is a non-heme iron enzyme, which oxidatively couples γ-glutamyl-cysteine and hercynine to produce a sulfoxide (**105** → **106**, Fig. 20A). The two substrates, hercynine (**105**) and γ-glutamyl-cysteine (**107**) are produced by two other enzymes in the biosynthetic gene cluster. EgtA catalyzes the coupling between glutamate and cysteine to produce γ-glutamyl-cysteine (**107**), while EgtD is the methyltransferase that adds three methyl groups to histidine amino group to produce hercynine (**105**). EgtC is a glutamine amidotransamidase homolog, which removes glutamate from the EgtB sulfoxide coupling product **106** to generate **108**. The last step of ergothioneine biosynthesis is catalyzed by a PLP-dependent C-S lyase (EgtE), whose activity has also been demonstrated *in vitro* (Irani et al., 2018; Song et al., 2015).

After the discovery of the mycobacterial EgtB gene, the functions of its homologs in fungi (*Neurospora crassa* (Bello, Barrera-Perez, Morin, & Epstein, 2012), *A. fumigates* (Sheridan et al., 2016), and the fission yeast, *Schizosaccharomyces pombe* (Pluskal, Ueno, & Yanagida, 2014)) have also been examined by genetic studies. Recent biochemical characterization of the Egt1 gene from fungal *N. crassa* indicated that this fungal Egt1 preferentially catalyzes the oxidative coupling between hercynine **105** and L-Cys to give hercyl-cysteine sulfoxide **108** (Fig. 20A) (Hu et al., 2014). These studies led to the elucidation of two aerobic ergothioneine biosynthetic pathways: the mycobacterial pathway (Seebeck, 2010) (EgtA – EgtE), and the fungal *N. crassa* pathway (Hu et al., 2014) (Egt1 and Egt2, Fig. 20A).Very recently, Seebeck analyzed the genomes of a green-sulfur bacterium *Chlorobium limicola* DSM 245 and identified Clim_1148 (EanA) as an EgtD homolog (Burn, Misson, Meury, & Seebeck, 2017). Analysis of EanA gene neighborhood revealed a putative rhodanese-like sulfur transferase (EanB, Clim_1149). In the presence of cysteine desulfurase IscS, EanB could mediate the sulfur transfer from cysteine to hercynine **105** to yield ergothioneine **109** (Fig. 20C). This discovery suggests that there exists at least three different ergothioneine biosynthetic pathways in nature: two aerobic (Fig. 20A) and one anaerobic (Fig. 20C) (Ruszczycky & Liu, 2017).

Ovothiol and its derivatives (**113**, Fig. 20B) were first isolated from the eggs or biological fluids of sea urchins, sea star, octopus, marine worms, and spiny scallop. Subsequent studies indicated that they are widely distributed among marine invertebrates, microalgae and proteobacteria, which has been systematically summarized by Castellano and Seebeck in one of their recent review papers (Castellano & Seebeck, 2018). In ovothiol, its thiol group is at the histidine side-chain δ-position and one of its most prominent properties is that the thiol group pK_a in ovothiol is 1.4, which is significantly lower than that of other natural thiols, including cysteine (p$K_{a,SH}$ = 8.4), glutathione (p$K_{a,SH}$ = 8.7), trypanothione (p$K_{a,SH}$ = 7.4), mycothiol (p$K_{a,SH}$ = 7.9), coenzyme A (p$K_{a,SH}$ = 9.8) or bacillithiol (p$K_{a,SH}$ = 8.0) (Castellano & Seebeck, 2018; Sharma, Van Laer, Messens, & Hamilton, 2016). This low pK_a makes ovothiol A reactive and could function as a nucleophile to reduce peroxides, while at the same time, it is oxidized to its disulfide form. The resulting ovothiol A disulfide can then be recycled back to ovothiol by glutathione and NAD(P)H dependent glutathione reductase system (Weaver & Rabenstein, 1995). Ovothiol biosynthetic studies were initiated about two decades ago, starting with partially fractionated cell lysate from an ovothiol-producing strain *Crithidia fasciculate*. Based on the biochemical data from these biochemical data, Steenkamp and co-workers proposed that this pathway (**104** → **111** → **112** → **113**) involves a sulfoxide synthase, a PLP-dependent lyase, and that the sulfur transfer process occurs prior to methylation (Fig. 20B) (Vogt, Spies, & Steenkamp, 2001; Steenkamp, Weldrick, & Spies, 1996). Recently, based on the sulfoxide synthase information, a non-heme iron enzyme OvoA in *Erwinia tasmaniensis* was identified, which catalyzes the oxidative coupling between L-His and L-Cys in ovothiol pathway (**104** → **111**, Fig. 20B) (Braunshausen & Seebeck, 2011; Chen et al., 2018; Hu et al., 2014; Mashabela & Seebeck, 2013; Song et al., 2013, 2014). OvoA is a bifunctional protein consisting of an unusual iron-dependent sulfoxide synthase at the *N*-terminus and an *S*-adenosyl methionine-dependent (SAM) methyltransferase at the *C*-terminus (Naowarojna, Huang, et al., 2018b).

When Seebeck and co-workers first reported their discovery of the OvoA enzyme, they proposed several possible catalytic mechanism (Braunshausen & Seebeck, 2011). Three computational EgtB mechanistic studies were also reported recently. $^{18}O_2$ and $H_2^{18}O$ (Wei, Siegbahn, & Liao, 2017; Faponle, Seebeck, & de Visser, 2017; Tian, Su, & Liu, 2018) Due to the similarity between EgtB- and OvoA-catalysis, three OvoA-mechanistic models were proposed and examined recently using unnatural amino acid containing OvoA (Fig. 21) (Chen et al., 2018, 2019). In pathways IA and IB, the first half of the reaction is the oxidation of cysteine to a sulfenic acid intermediate (**115**, Fig. 21). Based on the computational studies from Tian et al. on EgtB studies, Fe(IV)=O species **115** may deprotonate imidazole δ-position to generate an anion **116**, which nucleophilically attacks the sulfenic acid to produce sulfoxide **111**. In pathway IB, Fe(IV)=

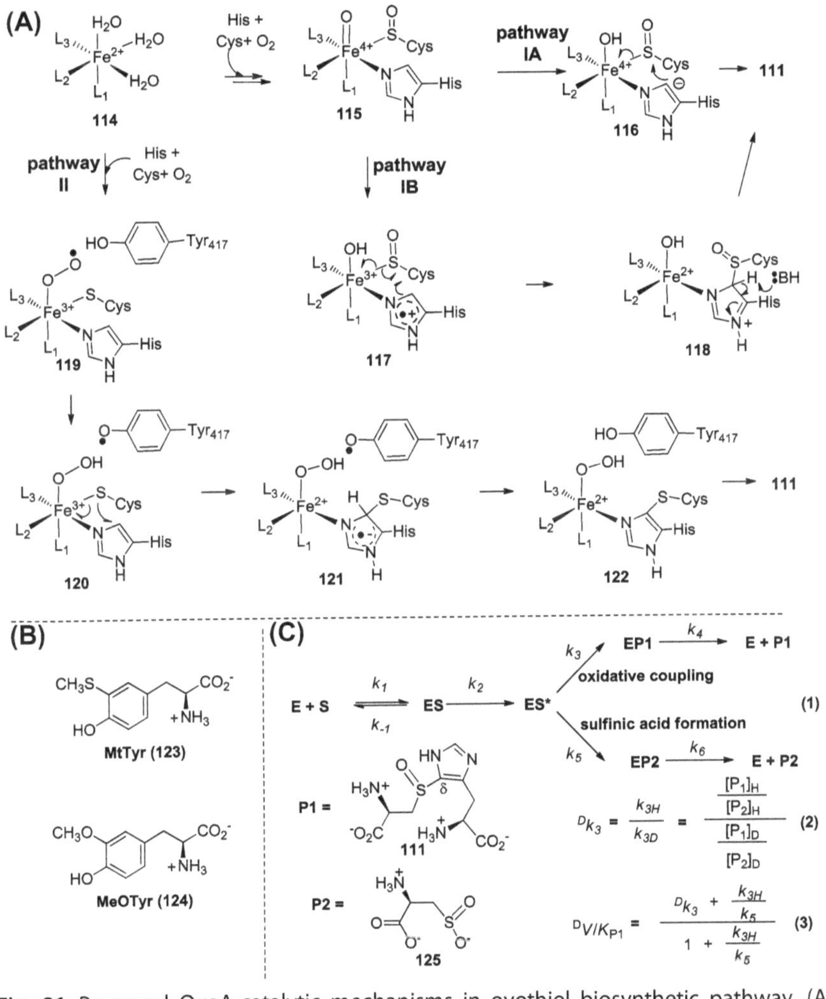

Fig. 21 Proposed OvoA-catalytic mechanisms in ovothiol biosynthetic pathway. (A) Proposed mechanisms; (B) Unnatural tyrosine used in OvoA-mechanistic studies; (C) Kinetic isotope sensitive branching method using in OvoA-mechanistic studies.

O oxidizes the imidazole side-chain to produce a cation radical **117** and it combines with sulfenic acid to form the C-S bond in **118**. An active site tyrosine functions as the base (species **118**, Fig. 21) (Tian et al., 2018). Wei et al. proposed that this deprotonation step is the rate-limiting step in EgtB-catalysis (Wei et al., 2017). Pathway II was examined by Faponle et al. They suggested that an active site tyrosine is part of a proton-coupled electron transfer process and oxidative C-S bond formation is the starting point for this reaction (**120** → **121**, Fig. 21) (Faponle et al., 2017). Faponle et al. also suggested that the **120** → **121** reaction is the rate-limiting step and inverse deuterium isotopic effect is predicated if the δ-hydrogen is replaced with deuterium (secondary deuterium isotope effect). Recently, based on an OvoA homology model, we replaced a proposed active site tyrosine residue in OvoA (Tyr417) by two unnatural amino acids, 2-amino-3-(4-hydroxy-3-(methylthio) phenyl) propanoic acid (MtTyr, **123**, Fig. 21B) and 2-amino-3-(4-hydroxy-3-(methoxyl) phenyl) propanoic Acid (MeOTyr, **124**, Fig. 21B) through amber-suppressor mediated unnatural amino acid incorporation method (Chen et al., 2018, 2019). OvoA catalyzes two reactions simultaneously: the oxidative coupling of histidine and cysteine to produce **111** and the oxidation of cysteine to cysteine sulfinic acid (**125**, Fig. 21C). When we replaced Tyr417 by either MtTyr (**123**) or MeOTyr (**124**), we could modulate the two OvoA activities: oxidative coupling and sulfinic acid formation (Eq. 1, Fig. 21C). By taking advantage of the existence of two competing pathways in OvoA-catalysis (Eq. 1, Fig. 21C), we utilized kinetic isotope sensitive branching (Eq. 2, Fig. 21C) to measure isotope effect at one of the branching steps (Cleland, 1975; Korzekwa, Trager, & Gillette, 1989; Jones, Korzekwa, Rettie, & Trager, 1986) and substrate deuterium KIE on V/K ($^{D}V/K$) using intermolecular competition (Cleland et al., 2006; Colletto, Whitaker, & Larrosa, 2017; Gu & Zhang, 2013). In MeOTyr-containing OvoA, it has a different rate-limiting step relative to wild type OvoA-catalysis. In addition, MeOTyr OvoA-catalysis displays an inverse deuterium KIE when deuterium-labeled histidine was used as the substrate. These results are in line with proposed pathway II mechanistic model in Fig. 21A, in which the oxidative C-S bond formation precedes the sulfoxidation reaction in this four-electron oxidation process. Further studies on EgtB and Egt1 will shed light on whether these three enzymes follow a similar mechanism (Tian et al., 2018; Wei et al., 2017).

9. Conclusion

Non-heme iron enzymes are one of the widely used classes of tailoring enzymes in natural product biosynthesis (Gao et al., 2018; Hausinger, 2004; Martinez and Hausinger, 2015). In addition, non-heme iron enzymes are capable of catalyzing transformation as diverse as heme-containing enzymes (Solomon et al., 2000). Besides the most commonly observed hydroxylation reactions, many new types of transformations were uncovered, which play a key role in introducing the structural and functional diversities to natural products. In this paper, we summarized some of the very recent examples, including endoperoxidation, orthoester formation, cyclopropranation, and oxidative C-C and C-S bond formation reactions. Many of their mechanistic details are only beginning to emerge. Surely, knowledge gained on these systems will enrich our understanding on how nature fine-tunes the iron center to achieve such a wide range of diverse functions.

Acknowledgments

This work is supported in part by a grant from the National Science Foundation (CHE-1309148 to P.L.) and the Boston University Nano Center seeding grant.

References

Abu-Omar, M. M., Loaiza, A., & Hontzeas, N. (2005). Reaction mechanisms of mononuclear non-heme iron oxygenases. *Chemical Reviews, 105*(6), 2227–2252.

Allen, J. D., van Loevezijn, A., Lakhai, J. M., van der Valk, M., van Tellingen, O., Reid, G., et al. (2002). Potent and specific inhibition of the breast cancer resistance protein multidrug transporter in vitro and in mouse intestine by a novel analogue of fumitremorgin C. *Molecular Cancer Therapeutics, 1*(6), 417–425.

Arduini, A., Eddy, L., & Hochstein, P. (1990). The reduction of ferryl myoglobin by ergothioneine - a novel function for ergothioneine. *Archives of Biochemistry and Biophysics, 281*(1), 41–43.

Arenz, S., Juette, M. F., Graf, M., Nguyen, F., Huter, P., Polikanov, Y. S., et al. (2016). Structures of the orthosomycin antibiotics avilamycin and evernimicin in complex with the bacterial 70S ribosome. *Proceedings of the National Academy of Sciences of the United States of America, 113*(27), 7527–7532.

Aruoma, O. I., Whiteman, M., England, T. G., & Halliwell, B. (1997). Antioxidant action of ergothioneine: Assessment of its ability to scavenge peroxynitrite. *Biochemical and Biophysical Research Communications, 231*(2), 389–391.

Bastard, J. P., Maachi, M., Van Nhieu, J. T., Jardel, C., Bruckert, E., Grimaldi, A., et al. (2002). Adipose tissue IL-6 content correlates with resistance to insulin activation of glucose uptake both in vivo and in vitro. *The Journal of Clinical Endocrinology & Metabolism, 87*(5), 2084–2089.

Bedirli, A., Sakrak, O., Muhtaroglu, S., Soyuer, I., Guler, I., Erdogan, A. R., et al. (2004). Ergothioneine pretreatment protects the liver from ischemia-reperfusion injury caused by increasing hepatic heat shock protein 70. *Journal of Surgical Research, 122*(1), 96–102.

Bello, M. H., Barrera-Perez, V., Morin, D., & Epstein, L. (2012). The *Neurospora crassa* mutant NcΔEgt-1 identifies an ergothioneine biosynthetic gene and demonstrates that ergothioneine enhances conidial survival and protects against peroxide toxicity during conidial germination. *Fungal Genetics and Biology, 49*(2), 160–172.

Blin, K., Wolf, T., Chevrette, M. G., Lu, X., Schwalen, C. J., Kautsar, S. A., et al. (2017). antiSMASH 4.0-improvements in chemistry prediction and gene cluster boundary identification. *Nucleic Acids Research, 45*(W1), W36–W41.

Blodgett, J. A. V., Thomas, P. M., Li, G. Y., Velasquez, J. E., van der Donk, W. A., Kelleher, N. L., et al. (2007). Unusual transformations in the biosynthesis of the antibiotic phosphinothricin tripeptide. *Nature Chemical Biology, 3*(8), 480–485.

Bohlmann, J., Meyer-Gauen, G., & Croteau, R. (1998). Plant terpenoid synthases: Molecular biology and phylogenetic analysis. *Proceedings of the National Academy of Sciences of the United States of America, 95*(8), 4126–4133.

Borthwick, A. D. (2012). 2,5-Diketopiperazines: Synthesis, reactions, Medicinal chemistry, and bioactive natural products. *Chemical Reviews, 112*(7), 3641–3716.

Brauer, A., Beck, P., Hintermann, L., & Groll, M. (2016). Structure of the dioxygenase AsqJ: Mechanistic insights into a one-pot multistep quinolone antibiotic biosynthesis. *Angewandte Chemie International Edition, 55*(1), 422–426.

Braunshausen, A., & Seebeck, F. P. (2011). Identification and characterization of the first ovothiol biosynthetic enzyme. *Journal of the American Chemical Society, 133*(6), 1757–1759.

Briggs, I. (1972). Ergothioneine in central nervous-system. *Journal of Neurochemistry, 19*(1), 27–&.

Buongiorno, D., & Straganz, G. D. (2013). Structure and function of atypically coordinated enzymatic mononuclear non-heme-Fe(II) centers. *Coordination Chemistry Reviews, 257*(2), 541–563.

Burn, R., Misson, L., Meury, M., & Seebeck, F. P. (2017). Anaerobic origin of ergothioneine. *Angewandte Chemie International Edition, 56*(41), 12508–12511.

Carter, A. P., Clemons, W. M., Brodersen, D. E., Morgan-Warren, R. J., Wimberly, B. T., & Ramakrishnan, V. (2000). Functional insights from the structure of the 30S ribosomal subunit and its interactions with antibiotics. *Nature, 407*(6802), 340–348.

Casteel, D. A. (1992). Peroxy natural products. *Natural Product Reports, 9*(4), 289–312.

Casteel, D. A. (1999). Peroxy natural products. *Natural Product Reports, 16*(1), 55–73.

Castellano, I., & Seebeck, F. P. (2018). On ovothiol biosynthesis and biological roles: From life in the ocean to therapeutic potential. *Natural Product Reports, 35*(12), 1241–1250.

Chang, W. C., Dey, M., Liu, P. H., Mansoorabadi, S. O., Moon, S. J., Zhao, Z. B. K., et al. (2013). Mechanistic studies of an unprecedented enzyme-catalysed 1,2-phosphono-migration reaction. *Nature, 496*(7443), 114–118.

Chang, W. C., Guo, Y., Wang, C., Butch, S. E., Rosenzweig, A. C., Boal, A. K., et al. (2014). Mechanism of the C5 stereoinversion reaction in the biosynthesis of carbapenem antibiotics. *Science, 343*(6175), 1140–1144.

Chang, W.-C., Li, J., Lee, J. L., Cronican, A. A., & Guo, Y. (2016). Mechanistic investigation of a non-heme iron enzyme catalyzed epoxidation in (−)-4′-methoxycyclopenin biosynthesis. *Journal of the American Chemical Society, 138*(33), 10390–10393.

Chau, R., Kalaitzis, J. A., & Neilan, B. A. (2011). On the origins and biosynthesis of tetrodotoxin. *Aquatic Toxicology, 104*(1–2), 61–72.

Chau, R., Kalaitzis, J. A., Wood, S. A., & Neilan, B. A. (2013). Diversity and biosynthetic potential of culturable microbes associated with toxic marine animals. *Marine Drugs, 11*(8), 2695–2712.

Cheah, I. K., & Halliwell, B. (2012). Ergothioneine; antioxidant potential, physiological function and role in disease. *Biochimica et Biophysica Acta, 1822*(5), 784–793.

Cheng, J. Z., Coyle, C. M., Panaccione, D. G., & O'Connor, S. E. (2010). A role for old yellow enzyme in ergot alkaloid biosynthesis. *Journal of the American Chemical Society, 132*(6), 1776−+.

Chen, L., Naowarojna, N., Chen, B., Xu, M. L., Quill, M., Wang, J. Y., et al. (2019). Mechanistic studies of a nonheme iron enzyme OvoA in ovothiol biosynthesis using a tyrosine analogue, 2-amino-3-(4-hydroxy-3-(methoxyl) phenyl) propanoic acid (MeOTyr). *ACS Catalysis, 9*(1), 253−258.

Chen, L., Naowarojna, N., Song, H., Wang, S., Wang, J., Deng, Z., et al. (2018). Use of a tyrosine analogue to modulate the two activities of a nonheme iron enzyme OvoA in ovothiol biosynthesis, cysteine oxidation versus oxidative C−S bond formation. *Journal of the American Chemical Society, 140*(13), 4604−4612.

Cimermancic, P., Medema, M. H., Claesen, J., Kurita, K., Brown, L. C. W., Mavrommatis, K., et al. (2014). Insights into secondary metabolism from a global analysis of prokaryotic biosynthetic gene clusters. *Cell, 158*(2), 412−421.

Cleland, W. W. (1975). Partition analysis and the concept of net rate Constants as tools in enzyme kinetics. *Biochemistry, 14*(14), 3220−3224.

Cleland, W. W. (2006). Chapter 37: Enzyme mechanisms from isotope effects. In A. Kohen, & H.-H. Limbach (Eds.), *Isotope effects in Chemistry and Biology* (pp. 915−930). CRC Taylor & Francis.

Clifton, I. J., Doan, L. X., Sleeman, M. C., Topf, M., Suzuki, H., Wilmouth, R. C., et al. (2003). Crystal structure of carbapenem synthase (CarC). *Journal of Biological Chemistry, 278*(23), 20843−20850.

Clifton, I. J., McDonough, M. A., Ehrismann, D., Kershaw, N. J., Granatino, N., & Schofield, C. J. (2006). Structural studies on 2-oxoglutarate oxygenases and related double-stranded β-helix fold proteins. *Journal of Inorganic Biochemistry, 100*(4), 644−669.

Colletto, C., Whitaker, D., & Larrosa, I. (2017). Determination of H-2 KIEs from competition experiments: Increased Accuracy via isotopic enrichment. *Topics in Catalysis, 60*(8), 589−593.

Costas, M., Mehn, M. P., Jensen, M. P., & Que, L., Jr. (2004). Dioxygen activation at mononuclear nonheme iron active sites: Enzymes, models, and intermediates. *Chemical Reviews, 104*(2), 939−986.

Crossland, J., Mitchell, J. F., & Woodruff, G. N. (1966). Presence of ergothioneine in central nervous system and its probable identity with cerebellar factor. *The Journal of Physiology, 182*(2), 427−&.

D'Onofrio, N., Servillo, L., Giovane, A., Casale, R., Vitiello, M., Marfella, R., et al. (2016). Ergothioneine oxidation in the protection against high-glucose induced endothelial senescence: Involvement of SIRT1 and SIRT6. *Free Radical Biology & Medicine, 96*, 211−222.

Edwards, D. J., & Gerwick, W. H. (2004). Lyngbyatoxin biosynthesis: Sequence of biosynthetic gene cluster and identification of a novel aromatic prenyltransferase. *Journal of the American Chemical Society, 126*(37), 11432−11433.

Eustice, D. C., & Wilhelm, J. M. (1984). Mechanisms of action of aminoglycoside antibiotics in eucaryotic protein synthesis. *Antimicrobial Agents and Chemotherapy, 26*(1), 53−60.

Fahey, R. C. (2001). Novel thiols of prokaryotes. *Annual Review of Microbiology, 55*, 333−356.

Faponle, A. S., Seebeck, F. P., & de Visser, S. P. (2017). Sulfoxide synthase versus cysteine dioxygenase reactivity in a non-heme iron enzyme. *Journal of the American Chemical Society, 139*(27), 9259−9270.

Fayos, J., Lokensgard, D., & Clardy, J. (1974). Structure of verruculogen, a tremor Producing peroxide from *Penicillium verruculosum*. *Journal of the American Chemical Society, 96*(21), 6785−6787.

Franzoni, F., Colognato, R., Galetta, F., Laurenza, I., Barsotti, M., Di Stefano, R., et al. (2006). An in vitro study on the free radical scavenging capacity of ergothioneine:

Comparison with reduced glutathione, uric acid and trolox. *Biomedicine & Pharmacotherapy, 60*(8), 453–457.

Furuta, T., Koike, M., & Abe, M. (1982). Isolation of cycloclavine from the culture Broth of *Aspergillus japonicus saito. Agricultural & Biological Chemistry, 46*(7), 1921–1922.

Galonić, D. P., Barr, E. W., Walsh, C. T., Bollinger, J. M., & Krebs, C. (2007). Two interconverting Fe (IV) intermediates in aliphatic chlorination by the halogenase CytC3. *Nature Chemical Biology, 3*(2), 113–116.

Gao, S. S., Naowarojna, N., Cheng, R., Liu, X., & Liu, P. (2018). Recent examples of alpha-ketoglutarate-dependent mononuclear non-haem iron enzymes in natural product biosyntheses. *Natural Product Reports, 35*(8), 792–837.

Gaudelli, N. M., & Townsend, C. A. (2014). Epimerization and substrate gating by a TE domain in beta-lactam antibiotic biosynthesis. *Nature Chemical Biology, 10*(4), 251–258.

Grigat, S., Harlfinger, S., Pal, S., Striebinger, R., Golz, S., Geerts, A., et al. (2007). Probing the substrate specificity of the ergothioneine transporter with methimazole, hercynine, and organic cations. *Biochemical Pharmacology, 74*(2), 309–316.

Grobe, N., Zhang, B., Fisinger, U., Kutchan, T. M., Zenk, M. H., & Guengerich, F. P. (2009). Mammalian cytochrome P450 enzymes catalyze the phenol-coupling step in endogenous morphine biosynthesis. *Journal of Biological Chemistry, 284*(36), 24425–24431.

Grundemann, D., Harlfinger, S., Golz, S., Geerts, A., Lazar, A., Berkels, R., et al. (2005). Discovery of the ergothioneine transporter. *Proceedings of the National Academy of Sciences of the United States of America, 102*(14), 5256–5261.

Grundmann, A., Kuznetsova, T., Afiyatullov, S. S., & Li, S. M. (2008). FtmPT2, an *N*-prenyltransferase from *Aspergillus fumigatus*, catalyses the last step in the biosynthesis of fumitremorgin B. *ChemBioChem, 9*(13), 2059–2063.

Grundmann, A., & Li, S. M. (2005). Overproduction, purification and characterization of FtmPT1, a brevianamide F prenyltransferase from Aspergillus fumigatus. *Microbiology, 151*(Pt 7), 2199–2207.

Gu, H., & Zhang, S. M. (2013). Advances in kinetic isotope effect Measurement Techniques for enzyme mechanism study. *Molecules, 18*(8), 9278–9292.

Hanauske-Abel, H., & Günzler, V. (1982). A stereochemical concept for the catalytic mechanism of prolylhydroxylase: Applicability to classification and design of inhibitors. *Journal of Theoretical Biology, 94*(2), 421–455.

Hand, C. E., & Honek, J. F. (2005). Biological chemistry of naturally occurring thiols of microbial and marine origin. *Journal of Natural Products, 68*(2), 293–308.

Hangasky, J. A., Taabazuing, C. Y., Valliere, M. A., & Knapp, M. J. (2013). Imposing function down a (cupin)-barrel: Secondary structure and metal stereochemistry in the αKG-dependent oxygenases. *Metallomics, 5*(4), 287–301.

Hartman, P. E. (1990). Ergothioneine as antioxidant. *Methods in Enzymology, 186*, 310–318.

Hausinger, R. P. (2004). Fe(II)/alpha-ketoglutarate-dependent hydroxylases and related enzymes. *Critical Reviews in Biochemistry and Molecular Biology, 39*(1), 21–68.

Hegg, E. L., & Que, L. (1997). The 2-His-1-carboxylate facial triad - an emerging structural motif in mononuclear non-heme iron(II) enzymes. *European Journal of Biochemistry, 250*(3), 625–629.

Higgins, L. J., Yan, F., Liu, P. H., Liu, H. W., & Drennan, C. L. (2005). Structural insight into antibiotic fosfomycin biosynthesis by a mononuclear iron enzyme. *Nature, 437*(7060), 838–844.

Hoffart, L. M., Barr, E. W., Guyer, R. B., Bollinger, J. M., & Krebs, C. (2006). Direct spectroscopic detection of a CH-cleaving high-spin Fe (IV) complex in a prolyl-4-hydroxylase. *Proceedings of the National Academy of Sciences of the United States of America, 103*(40), 14738–14743.

Hofmann, C., Boll, R., Heitmann, B., Hauser, G., Durr, C., Frerich, A., et al. (2005). Genes encoding enzymes responsible for biosynthesis of L-lyxose and attachment of eurekanate during avilamycin biosynthesis. *Chemistry & Biology, 12*(10), 1137–1143.

Horak, R. M., & Vleggaar, R. (1987). Biosynthesis of verruculogen, a tremorgenic metabolite of *Penicillium verruculosum*: Stereochemical course of peroxide ring formation. *Journal of the Chemical Society, Chemical Communications, 20*, 1568–1570.

Huang, X., & Groves, J. T. (2017). Beyond ferryl-mediated hydroxylation: 40 years of the rebound mechanism and C–H activation. *Journal of Biological Inorganic Chemistry, 22*(2–3), 185–207.

Hu, W., Song, H., Sae Her, A., Bak, D. W., Naowarojna, N., Elliott, S. J., et al. (2014). Bioinformatic and biochemical characterizations of C-S bond formation and cleavage enzymes in the fungus *Neurospora crassa* ergothioneine biosynthetic pathway. *Organic Letters, 16*(20), 5382–5385.

Ikezawa, N., Iwasa, K., & Sato, F. (2008). Molecular cloning and characterization of CYP80G2, a cytochrome p450 that catalyzes an intramolecular C - C phenol coupling of (S)-reticuline in magnoflorine biosynthesis, from cultured Coptis japonica cells. *Journal of Biological Chemistry, 283*(14), 8810–8821.

Irani, S., Naowarojna, N., Tang, Y., Kathuria, K. R., Wang, S., Dhembi, A., et al. (2018). Snapshots of CS cleavage in Egt2 reveals substrate specificity and reaction mechanism. *Cell Chemical Biology, 25*(5), 519–529.

Ishikawa, K., Sato, F., Itabashi, T., Wachi, H., Takeda, H., Wakana, D., et al. (2016). Asnovolins A-G, spiromeroterpenoids isolated from the fungus *Aspergillus novofumigatus,* and suppression of fibronectin expression by asnovolin E. *Journal of Natural Products, 79*(9), 2167–2174.

Jain, H. D., Zhang, C., Zhou, S., Zhou, H., Ma, J., Liu, X., et al. (2008). Synthesis and structure-activity relationship studies on tryprostatin A, an inhibitor of breast cancer resistance protein. *Bioorganic & Medicinal Chemistry, 16*(8), 4626–4651.

Jakubczyk, D., Caputi, L., Hatsch, A., Nielsen, C. A. F., Diefenbacher, M., Klein, J., et al. (2015). Discovery and reconstitution of the cycloclavine biosynthetic pathway-enzymatic formation of a cyclopropyl group. *Angewandte Chemie International Edition, 54*(17), 5117–5121.

Jakubczyk, D., Caputi, L., Stevenson, C. E. M., Lawson, D. M., & O'Connor, S. E. (2016). Structural characterization of EasH (Aspergillus japonicus) - an oxidase involved in cycloclavine biosynthesis. *Chemical Communications, 52*(99), 14306–14309.

Jakubczyk, D., Cheng, J. Z., & O'Connor, S. E. (2014). Biosynthesis of the ergot alkaloids. *Natural Product Reports, 31*(10), 1328–1338.

Jal, S., & Khora, S. S. (2015). An overview on the origin and production of tetrodotoxin, a potent neurotoxin. *Journal of Applied Microbiology, 119*(4), 907–916.

Ji, J. N., & Chen, S. L. (2018). Asymmetric abstraction of two chemically-equivalent methylene hydrogens: Significant enantioselectivity of endoperoxide presented by fumitremorgin B endoperoxidase. *Physical Chemistry Chemical Physics, 20*(41), 26500–26505.

Jin, F. H., Maruyama, J., Juvvadi, P. R., Arioka, M., & Kitamoto, K. (2004). Development of a novel quadruple auxotrophic host transformation system by argB gene disruption using adeA gene and exploiting adenine auxotrophy in Aspergillus oryzae. *FEMS Microbiology Letters, 239*(1), 79–85.

Jones, J. P., Korzekwa, K. R., Rettie, A. E., & Trager, W. F. (1986). Branching and its effect on the observed intramolecular isotope effects in cytochrome-P450 catalyzed-reactions -a new method for the Estimation of Intrinsic isotope effects. *Journal of the American Chemical Society, 108*(22), 7074–7078.

Kal, S., & Que, L. (2017). Dioxygen activation by nonheme iron enzymes with the 2-His-1-carboxylate facial triad that generate high-valent oxoiron oxidants. *Journal of Biological Inorganic Chemistry, 22*(2–3), 339–365.

Kaneko, I., Takeuchi, Y., Yamaoka, Y., Tanaka, Y., Fukuda, T., Fukumori, Y., et al. (1980). Quantitative-determination of ergothioneine in plasma and tissues by TLC-densitometry. *Chemical and Pharmaceutical Bulletin, 28*(10), 3093–3097.
Kato, Y., Kubo, Y., Iwata, D., Kato, S., Sudo, T., Sugiura, T., et al. (2010). Gene knockout and metabolome analysis of carnitine/organic cation transporter OCTN1. *Pharmaceutical Research, 27*(5), 832–840.
Kato, N., Suzuki, H., Takagi, H., Asami, Y., Kakeya, H., Uramoto, M., et al. (2009). Identification of cytochrome P450s required for fumitremorgin biosynthesis in Aspergillus fumigatus. *ChemBioChem, 10*(5), 920–928.
Kato, N., Suzuki, H., Takagi, H., Uramoto, M., Takahashi, S., & Osada, H. (2011). Gene disruption and biochemical characterization of verruculogen synthase of Aspergillus fumigatus. *ChemBioChem, 12*(5), 711–714.
Kato, H., Yoshida, T., Tokue, T., Nojiri, Y., Hirota, H., Ohta, T., et al. (2007). Prenylated indole alkaloids isolated from a marine-derived fungus, Aspergillus sp. *Angewandte Chemie International Edition, 46*(13), 2254–2256.
Khaldi, N., Seifuddin, F. T., Turner, G., Haft, D., Nierman, W. C., Wolfe, K. H., et al. (2010). SMURF: Genomic mapping of fungal secondary metabolite clusters. *Fungal Genetics and Biology, 47*(9), 736–741.
Khoufache, K., Puel, O., Loiseau, N., Delaforge, M., Rivollet, D., Coste, A., et al. (2007). Verruculogen associated with Aspergillus fumigatus hyphae and conidia modifies the electrophysiological properties of human nasal epithelial cells. *BMC Microbiology, 7.*
Kodato, S., Nakagawa, M., Hongu, M., Kawate, T., & Hino, T. (1988). Total synthesis of (+)-fumitremorgin B, its epimeric isomers, and demethoxy derivatives. *Tetrahedron, 44*(2), 359–377.
Komaki, H., Ichikawa, N., Oguchi, A., Hamada, M., Tamura, T., Fujita, N., et al. (2017). Genome analysis-based reclassification of Streptomyces endus and Streptomyces sporocinereus as later heterotypic synonyms of Streptomyces hygroscopicus subsp. hygroscopicus. *International Journal of Systematic and Evolutionary Microbiology, 67*(2), 343–345.
Korzekwa, K. R., Trager, W. F., & Gillette, J. R. (1989). Theory for the observed isotope effects from enzymatic systems that form multiple products via branched reaction pathways - cytochrome-P450. *Biochemistry, 28*(23), 9012–9018.
Kovaleva, E. G., & Lipscomb, J. D. (2008). Versatility of biological non-heme Fe(II) centers in oxygen activation reactions. *Nature Chemical Biology, 4*(3), 186–193.
Krebs, C., Galonic Fujimori, D., Walsh, C. T., & Bollinger, J. M., Jr. (2007). Non-heme Fe (IV)–oxo intermediates. *Accounts of Chemical Research, 40*(7), 484–492.
Krishnamurthy, P., & Schuetz, J. D. (2006). Role of ABCG2/BCRP in biology and medicine. *Annual Review of Pharmacology and Toxicology, 46*, 381–410.
Kryatov, S. V., Rybak-Akimova, E. V., & Schindler, S. (2005). Kinetics and mechanisms of formation and reactivity of non-heme iron oxygen intermediates. *Chemical Reviews, 105*(6), 2175–2226.
Leone, E., & Mann, T. (1951). Ergothioneine in the seminal vesicle secretion. *Nature, 168*(4266), 205–206.
Leung, E., Hong, J. W., Fraser, A. G., Merriman, T. R., Vishnu, P., & Krissansen, G. W. (2006). Polymorphisms in the organic cation transporter genes SLC22A4 and SLC22A5 and Crohn's disease in a New Zealand Caucasian cohort. *Immunology & Cell Biology, 84*(2), 233–236.
Levesque, F., & Seeberger, P. H. (2012). Continuous-flow synthesis of the anti-malaria drug artemisinin. *Angewandte Chemie International Edition, 51*(7), 1706–1709.
Li, J., van Belkum, M. J., & Vederas, J. C. (2012). Functional characterization of recombinant hyoscyamine 6 beta-hydroxylase from Atropa belladonna. *Bioorganic & Medicinal Chemistry, 20*(14), 4356–4363.

Li, S., Zhang, J., Liu, Y., Sun, G., Deng, Z., & Sun, Y. (2018). Direct genetic and enzymatic evidence for oxidative cyclization in hygromycin B biosynthesis. *ACS Chemical Biology, 13*(8), 2203–2210.

Li, S. M. (2010). Prenylated indole derivatives from fungi: Structure diversity, biological activities, biosynthesis and chemoenzymatic synthesis. *Natural Product Reports, 27*(1), 57–78.

Liang, P. H., Ko, T. P., & Wang, A. H. (2002). Structure, mechanism and function of prenyltransferases. *European Journal of Biochemistry, 269*(14), 3339–3354.

Libby, P., Ridker, P. M., & Hansson, G. K. (2011). Progress and challenges in translating the biology of atherosclerosis. *Nature, 473*(7347), 317–325.

Liu, A., Ho, R. Y., Que, L., Jr., Ryle, M. J., Phinney, B. S., & Hausinger, R. P. (2001). Alternative reactivity of an alpha-ketoglutarate-dependent iron(II) oxygenase: Enzyme self-hydroxylation. *Journal of the American Chemical Society, 123*(21), 5126–5127.

Liu, P., Mehn, M. P., Yan, F., Zhao, Z., Que, L., Jr., & Liu, H. W. (2004). Oxygenase activity in the self-hydroxylation of (s)-2-hydroxypropylphosphonic acid epoxidase involved in fosfomycin biosynthesis. *Journal of the American Chemical Society, 126*(33), 10306–10312.

Liu, P. H., Murakami, K., Seki, T., He, X. M., Yeung, S. M., Kuzuyama, T., et al. (2001). Protein purification and function assignment of the epoxidase catalyzing the formation of fosfomycin. *Journal of the American Chemical Society, 123*(19), 4619–4620.

Lundberg, M., & Borowski, T. (2013). Oxoferryl species in mononuclear non-heme iron enzymes: Biosynthesis, properties and reactivity from a theoretical perspective. *Coordination Chemistry Reviews, 257*(1), 277–289.

Maiya, S., Grundmann, A., Li, S. M., & Turner, G. (2006). The fumitremorgin gene cluster of *Aspergillus fumigatus*: Identification of a gene encoding brevianamide F synthetase. *ChemBioChem, 7*(7), 1062–1069.

Marnett, L. J. (2000). Cyclooxygenase mechanisms. *Current Opinion in Chemical Biology, 4*(5), 545–552.

Martin, K. R. (2010). The bioactive agent ergothioneine, a key component of dietary mushrooms, inhibits monocyte binding to endothelial cells characteristic of early cardiovascular disease. *Journal of Medicinal Food, 13*(6), 1340–1346.

Martinez, S., & Hausinger, R. P. (2015). Catalytic mechanisms of Fe(II)- and 2-oxoglutarate-dependent oxygenases. *Journal of Biological Chemistry, 290*(34), 20702–20711.

Mashabela, G. T. M., & Seebeck, F. P. (2013). Substrate specificity of an oxygen dependent sulfoxide synthase in ovothiol biosynthesis. *Chemical Communications, 49*(70), 7714–7716.

Matsuda, Y., & Abe, I. (2016). Biosynthesis of fungal meroterpenoids. *Natural Product Reports, 33*(1), 26–53.

Matsuda, Y., Awakawa, T., Wakimoto, T., & Abe, I. (2013). Spiro-ring formation is catalyzed by a multifunctional dioxygenase in austinol biosynthesis. *Journal of the American Chemical Society, 135*(30), 10962–10965.

Matsuda, Y., Bai, T., Phippen, C. B. W., Nodvig, C. S., Kjaerbolling, I., Vesth, T. C., et al. (2018). Novofumigatonin biosynthesis involves a non-heme iron-dependent endoperoxide isomerase for orthoester formation. *Nature Communications, 9*(1), 2587.

Matsuda, Y., Iwabuchi, T., Fujimoto, T., Awakawa, T., Nakashima, Y., Mori, T., et al. (2016). Discovery of key dioxygenases that diverged the paraherquonin and acetoxydehydroaustin pathways in *Penicillium brasilianum*. *Journal of the American Chemical Society, 138*(38), 12671–12677.

Matsuda, Y., Wakimoto, T., Mori, T., Awakawa, T., & Abe, I. (2014). Complete biosynthetic pathway of anditomin: Nature's sophisticated synthetic route to a complex fungal meroterpenoid. *Journal of the American Chemical Society, 136*(43), 15326–15336.

Matthews, M. L., Krest, C. M., Barr, E. W., Vaillancourt, F. H., Walsh, C. T., Green, M. T., et al. (2009). Substrate-triggered formation and remarkable stability of the C– H bond-

cleaving chloroferryl intermediate in the aliphatic halogenase, SyrB2. *Biochemistry, 48*(20), 4331–4343.
Matthews, M. L., Neumann, C. S., Miles, L. A., Grove, T. L., Booker, S. J., Krebs, C., et al. (2009). Substrate positioning controls the partition between halogenation and hydroxylation in the aliphatic halogenase, SyrB2. *Proceedings of the National Academy of Sciences of the United States of America, 106*(42), 17723–17728.
Matuschek, M., Wallwey, C., Xie, X. L., & Li, S. M. (2011). New insights into ergot alkaloid biosynthesis in Claviceps purpurea: An agroclavine synthase EasG catalyses, via a nonenzymatic adduct with reduced glutathione, the conversion of chanoclavine-I aldehyde to agroclavine. *Organic and Biomolecular Chemistry, 9*(11), 4328–4335.
McCulloch, K. M., McCranie, E. K., Smith, J. A., Sarwar, M., Mathieu, J. L., Gitschlag, B. L., et al. (2015). Oxidative cyclizations in orthosomycin biosynthesis expand the known chemistry of an oxygenase superfamily. *Proceedings of the National Academy of Sciences of the United States of America, 112*(37), 11547–11552.
McDonough, M. A., Loenarz, C., Chowdhury, R., Clifton, I. J., & Schofield, C. J. (2010). Structural studies on human 2-oxoglutarate dependent oxygenases. *Current Opinion in Structural Biology, 20*(6), 659–672.
McGaha, S. M., & Champney, W. S. (2007). Hygromycin B inhibition of protein synthesis and ribosome biogenesis in *Escherichia coli*. *Antimicrobial Agents and Chemotherapy, 51*(2), 591–596.
Melville, D. B., Horner, W. H., & Lubschez, R. (1954). Tissue ergothioneine. *Journal of Biological Chemistry, 206*(1), 221–228.
Moncaster, J. A., Walsh, D. T., Gentleman, S. M., Jen, L. S., & Aruoma, O. I. (2002). Ergothioneine treatment protects neurons against N-methyl-D-aspartate excitotoxicity in an in vivo rat retinal model. *Neuroscience Letters, 328*(1), 55–59.
Morris, G. M., Goodsell, D. S., Halliday, R. S., Huey, R., Hart, W. E., Belew, R. K., et al. (1998). Automated docking using a Lamarckian genetic algorithm and an empirical binding free energy function. *Journal of Computational Chemistry, 19*(14), 1639–1662.
Mundt, K., Wollinsky, B., Ruan, H. L., Zhu, T. J., & Li, S. M. (2012). Identification of the verruculogen prenyltransferase FtmPT3 by a combination of chemical, bioinformatic and biochemical approaches. *ChemBioChem, 13*(17), 2583–2592.
Murphy, M. P., Holmgren, A., Larsson, N. G., Halliwell, B., Chang, C. J., Kalyanaraman, B., et al. (2011). Unraveling the biological roles of reactive oxygen species. *Cell Metabolism, 13*(4), 361–366.
Nakamura, T., Yoshida, K., Yabuuchi, H., Maeda, T., & Tamai, I. (2008). Functional characterization of ergothioneine transport by rat organic cation/carnitine transporter OCTN1 (slc22a4). *Biological and Pharmaceutical Bulletin, 31*(8), 1580–1584.
Naowarojna, N., Cheng, R., Chen, L., Quill, M., Xu, M., Zhao, C., et al. (2018). Mini-review: Ergothioneine and ovothiol biosyntheses, an unprecedented trans-sulfur strategy in natural product biosynthesis. *Biochemistry, 57*(24), 3309–3325.
Naowarojna, N., Huang, P., Cai, Y. J., Song, H., Wu, L., Cheng, R. H., et al. (2018). In vitro reconstitution of the remaining steps in ovothiol A biosynthesis: C-S lyase and methyltransferase reactions. *Organic Letters, 20*(17), 5427–5430.
Nielsen, P. V., Beuchat, L. R., & Frisvad, J. C. (1988). Growth of and fumitremorgin production by Neosartorya fischeri as affected by temperature, light, and water activity. *Applied and Environmental Microbiology, 54*(6), 1504–1510.
Nierman, W. C., Pain, A., Anderson, M. J., Wortman, J. R., Kim, H. S., Arroyo, J., et al. (2005). Genomic sequence of the pathogenic and allergenic filamentous fungus Aspergillus fumigatus. *Nature, 438*(7071), 1151–1156.
Onaka, H., Asamizu, S., Igarashi, Y., Yoshida, R., & Furumai, T. (2005). Cytochrome P450 homolog is responsible for C-N bond formation between aglycone and deoxysugar in

the staurosporine biosynthesis of Streptomyces sp TP-A0274. *Bioscience, Biotechnology, and Biochemistry, 69*(9), 1753–1759.

Paddon, C. J., Westfall, P. J., Pitera, D. J., Benjamin, K., Fisher, K., McPhee, D., et al. (2013). High-level semi-synthetic production of the potent antimalarial artemisinin. *Nature, 496*(7446), 528–532.

Paul, B. D., & Snyder, S. H. (2010). The unusual amino acid L-ergothioneine is a physiologic cytoprotectant. *Cell Death & Differentiation, 17*(7), 1134–1140.

Peltekova, V. D., Wintle, R. F., Rubin, L. A., Amos, C. I., Huang, Q. Q., Gu, X. J., et al. (2004). Functional variants of OCTN cation transporter genes are associated with Crohn disease. *Nature Genetics, 36*(5), 471–475.

Pfister, P., Risch, M., Brodersen, D. E., & Bottger, E. C. (2003). Role of 16S rRNA Helix 44 in ribosomal resistance to hygromycin B. *Antimicrobial Agents and Chemotherapy, 47*(5), 1496–1502.

Phelan, R. M., & Townsend, C. A. (2013). Mechanistic insights into the bifunctional non-heme iron oxygenase carbapenem synthase by active site saturation mutagenesis. *Journal of the American Chemical Society, 135*(20), 7496–7502.

Pluskal, T., Ueno, M., & Yanagida, M. (2014). Genetic and metabolomic dissection of the ergothioneine and selenoneine biosynthetic pathway in the fission yeast, *S. pombe*, and construction of an overproduction system. *PLoS One, 9*(5), e97774.

Pojer, F., Wemakor, E., Kammerer, B., Chen, H., Walsh, C. T., Li, S. M., et al. (2003). CloQ, a prenyltransferase involved in clorobiocin biosynthesis. *Proceedings of the National Academy of Sciences of the United States of America, 100*(5), 2316–2321.

Price, J. C., Barr, E. W., Glass, T. E., Krebs, C., & Bollinger, J. M. (2003). Evidence for hydrogen abstraction from C1 of taurine by the high-spin Fe (IV) intermediate detected during oxygen activation by taurine: α-Ketoglutarate dioxygenase (TauD). *Journal of the American Chemical Society, 125*(43), 13008–13009.

Price, J. C., Barr, E. W., Tirupati, B., Bollinger, J. M., & Krebs, C. (2003). The first direct characterization of a high-valent iron intermediate in the reaction of an alpha-ketoglutarate-dependent dioxygenase: A high-spin Fe(IV) complex in taurine/alpha-ketoglutarate dioxygenase (TauD) from Escherichia coli. *Biochemistry, 42*(24), 7497–7508.

Que, L., Jr., & Ho, R. Y. (1996). Dioxygen activation by enzymes with mononuclear non-heme iron active sites. *Chemical Reviews, 96*(7), 2607–2624.

Rabindran, S. K., He, H. Y., Singh, M., Brown, E., Collins, K. I., Annable, T., et al. (1998). Reversal of a novel multidrug resistance mechanism in human colon carcinoma cells by fumitremorgin C. *Cancer Research, 58*(24), 5850–5858.

Rabindran, S. K., Ross, D. D., Doyle, L. A., Yang, W. D., & Greenberger, L. M. (2000). Fumitremorgin C reverses multidrug resistance in cells transfected with the breast cancer resistance protein. *Cancer Research, 60*(1), 47–50.

Ran, H. M., Wohlgemuth, V., Xie, X. L., & Li, S. M. (2018). A nonheme Fe-II/2-oxoglutarate-dependent oxygenase catalyzes a double bond migration within a dimethylallyl moiety accompanied by hydroxylation. *ACS Chemical Biology, 13*(10), 2949–2955.

Rank, C., Phipps, R. K., Harris, P., Fristrup, P., Larsen, T. O., & Gotfredsen, C. H. (2008). Novofumigatonin, a new orthoester meroterpenoid from Aspergillus novofumigatus. *Organic Letters, 10*(3), 401–404.

Ridker, P. M., Rifai, N., Stampfer, M. J., & Hennekens, C. H. (2000). Plasma concentration of interleukin-6 and the risk of future myocardial infarction among apparently healthy men. *Circulation, 101*(15), 1767–1772.

Rohde, J.-U., In, J.-H., Lim, M. H., Brennessel, W. W., Bukowski, M. R., Stubna, A., et al. (2003). Crystallographic and spectroscopic characterization of a nonheme Fe (IV)= O complex. *Science, 299*(5609), 1037–1039.

Rougee, M., Bensasson, R. V., Land, E. J., & Pariente, R. (1988). Deactivation of singlet molecular-oxygen by thiols and related-compounds, possible protectors against skin photosensitivity. *Photochemistry and Photobiology, 47*(4), 485–489.

Ruszczycky, M. W., & Liu, H. W. (2017). The surprising history of an antioxidant. *Nature, 551*(7678), 37–38.

Ryle, M. J., Liu, A., Muthukumaran, R. B., Ho, R. Y., Koehntop, K. D., McCracken, J., et al. (2003). O2- and alpha-ketoglutarate-dependent tyrosyl radical formation in TauD, an alpha-keto acid-dependent non-heme iron dioxygenase. *Biochemistry, 42*(7), 1854–1862.

Sacchettini, J. C., & Poulter, C. D. (1997). Creating isoprenoid diversity. *Science, 277*(5333), 1788–1789.

Sakrak, O., Kerem, M., Bedirli, A., Pasaoglu, H., Akyurek, N., Ofluoglu, E., et al. (2008). Ergothioneine modulates proinflammatory cytokines and heat shock protein 70 in mesenteric ischemia and reperfusion injury. *Journal of Surgical Research, 144*(1), 36–42.

Schofield, S. J., & Hausinger, R. P. (2015). *2-Oxoglutarate-dependent oxygenases*. Royal Society of Chemistry.

Scott, E. M., Duncan, I. W., & Ekstrand, V. (1963). Purification and properties of glutathione reductase of human erythrocytes. *Journal of Biological Chemistry, 238*, 3928–3933.

Seebeck, F. P. (2010). *In vitro* reconstitution of mycobacterial ergothioneine biosynthesis. *Journal of the American Chemical Society, 132*(19), 6632–6633.

Sharma, S. V., Van Laer, K., Messens, J., & Hamilton, C. J. (2016). Thiol redox and pKa properties of mycothiol, the predominant low-molecular-weight thiol cofactor in the actinomycetes. *ChemBioChem, 17*(18), 1689–1692.

Sheridan, K. J., Lechner, B. E., O'Keeffe, G., Keller, M. A., Werner, E. R., Lindner, H., et al. (2016). Ergothioneine biosynthesis and functionality in the opportunistic fungal pathogen, *Aspergillus fumigatus*. *Scientific Reports, 6*, 35306.

Shires, T. K., Brummel, M. C., Pulido, J. S., & Stegink, L. D. (1997). Ergothioneine distribution in bovine and porcine ocular tissues. *Comparative Biochemistry and Physiology - Part C: Toxicology & Pharmacology, 117*(1), 117–120.

Siitonen, V., Blauenburg, B., Kallio, P., Mantsala, P., & Metsa-Ketela, M. (2012). Discovery of a two-component monooxygenase SnoaW/SnoaL2 involved in nogalamycin biosynthesis. *Chemistry & Biology, 19*(5), 638–646.

Siitonen, V., Claesson, M., Patrikainen, P., Aromaa, M., Mantsala, P., Schneider, G., et al. (2012). Identification of late-stage glycosylation steps in the biosynthetic pathway of the anthracycline nogalamycin. *ChemBioChem, 13*(1), 120–128.

Siitonen, V., Selvaraj, B., Niiranen, L., Lindqvist, Y., Schneider, G., & Metsa-Ketela, M. (2016). Divergent non-heme iron enzymes in the nogalamycin biosynthetic pathway. *Proceedings of the National Academy of Sciences of the United States of America, 113*(19), 5251–5256.

Smith, W. L., Urade, Y., & Jakobsson, P. J. (2011). Enzymes of the cyclooxygenase pathways of prostanoid biosynthesis. *Chemical Reviews, 111*(10), 5821–5865.

Solomon, E. I., Brunold, T. C., Davis, M. I., Kemsley, J. N., Lee, S. K., Lehnert, N., et al. (2000). Geometric and electronic structure/function correlations in non-heme iron enzymes. *Chemical Reviews, 100*(1), 235–350.

Song, H., Her, A. S., Raso, F., Zhen, Z., Huo, Y., & Liu, P. (2014). Cysteine oxidation reactions catalyzed by a mononuclear non-heme iron enzyme (OvoA) in ovothiol biosynthesis. *Organic Letters, 16*(8), 2122–2125.

Song, H., Hu, W., Naowarojna, N., Her, A. S., Wang, S., Desai, R., et al. (2015). Mechanistic studies of a novel C-S lyase in ergothioneine biosynthesis: The involvement of a sulfenic acid intermediate. *Scientific Reports, 5*, 11870.

Song, H., Leninger, M., Lee, N., & Liu, P. (2013). Regioselectivity of the oxidative C–S bond formation in ergothioneine and ovothiol biosyntheses. *Organic Letters, 15*(18), 4854–4857.

Steenkamp, D. J., Weldrick, D., & Spies, H. S. C. (1996). Studies on the biosynthesis of ovothiol A - Identification of 4-mercaptohistidine as an intermediate. *European Journal of Biochemistry, 242*(3), 557–566.

Steffan, N., Grundmann, A., Afiyatullov, S., Ruan, H. L., & Li, S. M. (2009). FtmOx1, a non-heme Fe(II) and alpha-ketoglutarate-dependent dioxygenase, catalyses the endoperoxide formation of verruculogen in Aspergillus fumigatus. *Organic and Biomolecular Chemistry,* 7(19), 4082–4087.

Steffan, N., Grundmann, A., Yin, W. B., Kremer, A., & Li, S. M. (2009). Indole prenyltransferases from fungi: A new enzyme group with high potential for the production of prenylated indole derivatives. *Current Medicinal Chemistry, 16*(2), 218–231.

Stubbe, J., & van Der Donk, W. A. (1998). Protein radicals in enzyme catalysis. *Chemical Reviews, 98*(2), 705–762.

Tang, M. C., Zou, Y., Watanabe, K., Walsh, C. T., & Tang, Y. (2017). Oxidative cyclization in natural product biosynthesis. *Chemical Reviews, 117*(8), 5226–5333.

Taubert, D., Jung, N., Goeser, T., & Schomig, E. (2009). Increased ergothioneine tissue concentrations in carriers of the Crohn's disease risk-associated 503F variant of the organic cation transporter OCTN1. *Gut, 58*(2), 312–314.

Taubert, D., Lazar, A., Grimberg, G., Jung, N., Rubbert, A., Delank, K. S., et al. (2006). Association of rheumatoid arthritis with ergothioneine levels in red blood cells: A case control study. *Journal of Rheumatology, 33*(11), 2139–2145.

Thibodeaux, C. J., Chang, W. C., & Liu, H. W. (2012). Enzymatic chemistry of cyclopropane, epoxide, and aziridine biosynthesis. *Chemical Reviews, 112*(3), 1681–1709.

Tian, G., Su, H., & Liu, Y. J. (2018). Mechanism of sulfoxidation and C-S bond formation involved in the biosynthesis of ergothioneine catalyzed by ergothioneine synthase (EgtB). *ACS Catalysis, 8*(7), 5875–5889.

Tokuhiro, S., Yamada, R., Chang, X. T., Suzuki, A., Kochi, Y., Sawada, T., et al. (2003). An intronic SNP in a RUNX1 binding site of SLC22A4, encoding an organic cation transporter, is associated with rheumatoid arthritis. *Nature Genetics, 35*(4), 341–348.

Tsai, H. F., Wang, H., Gebler, J. C., Poulter, C. D., & Schardl, C. L. (1995). The Claviceps purpurea gene encoding dimethylallyltryptophan synthase, the committed step for ergot alkaloid biosynthesis. *Biochemical and Biophysical Research Communications, 216*(1), 119–125.

Tsukamoto, S., Kato, H., Samizo, M., Nojiri, Y., Onuki, H., Hirota, H., et al. (2008). Prenylated indole alkaloids isolated from a marine-derived Aspergillus sp. *Journal of Natural Products, 71*(12), 2064–2067.

Tudzynski, P., Holter, K., Correia, T., Arntz, C., Grammel, N., & Keller, U. (1999). Evidence for an ergot alkaloid gene cluster in Claviceps purpurea. *Molecular Genetics and Genomics, 261*(1), 133–141.

Unsold, I. A., & Li, S. M. (2005). Overproduction, purification and characterization of FgaPT2, a dimethylallyltryptophan synthase from Aspergillus fumigatus. *Microbiology, 151*(Pt 5), 1499–1505.

Valegård, K., van Scheltinga, A. C. T., Lloyd, M. D., Hara, T., Ramaswamy, S., Perrakis, A., et al. (1998). Structure of a cephalosporin synthase. *Nature, 394*(6695), 805–809.

Vey, J. L., Al-Mestarihi, A., Hu, Y., Funk, M. A., Bachmann, B. O., & Iverson, T. M. (2010). Structure and mechanism of ORF36, an amino sugar oxidizing enzyme in everninomicin biosynthesis. *Biochemistry, 49*(43), 9306–9317.

de Visser, S. P. (2010). Trends in substrate hydroxylation reactions by heme and nonheme iron(IV)-Oxo oxidants give correlations between Intrinsic properties of the oxidant with barrier Height. *Journal of the American Chemical Society, 132*(3), 1087–1097.

Vleggaar, R., Horak, R. M., & Maharaj, V. J. (1993). Stereochemical course of ring formation in fumitremorgin B and verruculogen, metabolites of *Penicillium verruculosum* - investigation into the loss of stereochemical integrity of the geminal methyl-groups. *Journal of the Chemical Society, Chemical Communications, 3*, 274–275.

Vogt, R. N., Spies, H. S. C., & Steenkamp, D. J. (2001). The biosynthesis of ovothiol A (N-1-methyl-4-mercaptohistidine) - identification of S-(4 '-L-histidyl)-L-cysteine sulfoxide as an intermediate and the products of the sulfoxide lyase reaction. *European Journal of Biochemistry, 268*(20), 5229–5241.

Wallwey, C., & Li, S. M. (2011). Ergot alkaloids: Structure diversity, biosynthetic gene clusters and functional proof of biosynthetic genes. *Natural Product Reports, 28*(3), 496–510.

Wang, X. Y., Su, H., & Liu, Y. J. (2017). Insights into the unprecedented epoxidation mechanism of fumitremorgin B endoperoxidase (FtmOx1) from Aspergillus fumigatus by QM/MM calculations. *Physical Chemistry Chemical Physics, 19*(11), 7668–7677.

Weaver, K. H., & Rabenstein, D. L. (1995). Thiol-disulfide exchange-reactions of ovothiol-a with glutathione. *Journal of Organic Chemistry, 60*(6), 1904–1907.

Wei, W. J., Siegbahn, P. E., & Liao, R. Z. (2017). Theoretical study of the mechanism of the non-heme iron enzyme EgtB. *Inorganic Chemistry, 56*(6), 3589–3599.

Weichold, V., Milbredt, D., & van Pée, K. H. (2016). Specific enzymatic halogenation—from the discovery of halogenated enzymes to their applications in vitro and in vivo. *Angewandte Chemie International Edition, 55*(22), 6374–6389.

Weitnauer, G., Gaisser, S., Kellenberger, L., Leadlay, P. F., & Bechthold, A. (2002). Analysis of a C-methyltransferase gene (aviG1) involved in avilamycin biosynthesis in Streptomyces viridochromogenes Tu57 and complementation of a Saccharopolyspora erythraea eryBIII mutant by aviG1. *Microbiology, 148*(Pt 2), 373–379.

Weitnauer, G., Muhlenweg, A., Trefzer, A., Hoffmeister, D., Sussmuth, R. D., Jung, G., et al. (2001). Biosynthesis of the orthosomycin antibiotic avilamycin A: Deductions from the molecular analysis of the avi biosynthetic gene cluster of Streptomyces viridochromogenes Tu57 and production of new antibiotics. *Chemistry & Biology, 8*(6), 569–581.

Williams, R. M., Stocking, E. M., & Sanz-Cervera, J. F. (2000). Biosynthesis of prenylated alkaloids derived from tryptophan. In F. J. Leeper, & J. C. Vederas (Eds.), *Top. Curr. Chem.* (Vol. 209, pp. 97–173).

Willingale, J., Perera, K. P., & Mantle, P. G. (1983). An intermediary role for the tremorgenic mycotoxin TR-2 in the biosynthesis of verruculogen. *Biochemical Journal, 214*(3), 991–993.

Woehlecke, H., Osada, H., Herrmann, A., & Lage, H. (2003). Reversal of breast cancer resistance protein-mediated drug resistance by tryprostatin A. *International Journal of Cancer, 107*(5), 721–728.

Wu, L. F., Meng, S., & Tang, G. L. (2016). Ferrous iron and alpha-ketoglutarate-dependent dioxygenases in the biosynthesis of microbial natural products. *Biochimica et Biophysica Acta - Proteins and Proteomics, 1864*(5), 453–470.

Yamazaki, M., Fujimoto, H., & Kawasaki, T. (1980). Chemistry of Tremorogenic metabolites .1. Fumitremorgin-a from Aspergillus-fumigatus. *Chemical and Pharmaceutical Bulletin, 28*(1), 245–254.

Yan, W., Song, H., Song, F., Guo, Y., Wu, C. H., Her, A. S., et al. (2015). Endoperoxide formation by an alpha-ketoglutarate-dependent mononuclear non-haem iron enzyme. *Nature, 527*(7579), 539–543.

Yaremenko, I. A., Vil, V. A., Demchuk, D. V., & Terent'ev, A. O. (2016). Rearrangements of organic peroxides and related processes. *Beilstein Journal of Organic Chemistry, 12*, 1647–1748.

Zhang, Z., Ren, J., Stammers, D. K., Baldwin, J. E., Harlos, K., & Schofield, C. J. (2000). Structural origins of the selectivity of the trifunctional oxygenase clavaminic acid synthase. *Nature Structural Biology, 7*(2), 127–133.

Zhao, L., Chang, W. C., Xiao, Y., Liu, H. W., & Liu, P. (2013). Methylerythritol phosphate pathway of isoprenoid biosynthesis. *Annual Review of Biochemistry, 82*, 497–530.

Zhao, W., Fan, A., Tarcz, S., Zhou, K., Yin, W. B., Liu, X. Q., et al. (2017). Mutation on Gly115 and Tyr205 of the cyclic dipeptide C2-prenyltransferase FtmPT1 increases its catalytic activity toward hydroxynaphthalenes. *Applied Microbiology and Biotechnology, 101*(5), 1989–1998.

Zhao, S., Smith, K. S., Deveau, A. M., Dieckhaus, C. M., Johnson, M. A., Macdonald, T. L., et al. (2002). Biological activity of the tryprostatins and their diastereomers on human carcinoma cell lines. *Journal of Medicinal Chemistry, 45*(8), 1559–1562.

CHAPTER TWO

Glutarate L-2-hydroxylase (CsiD/GlaH) is an archetype Fe(II)/2-oxoglutarate-dependent dioxygenase

Caitlyn Q Herr[a], Lee Macomber[b,1], Efthalia Kalliri[c,2] and Robert P. Hausinger[a,b,*]

[a]Department of Biochemistry and Molecular Biology, Michigan State University, East Lansing, MI, United States
[b]Department of Microbiology and Molecular Genetics, Michigan State University, East Lansing, MI, United States
[c]Department of Chemistry, Michigan State University, East Lansing, MI, United States
*Corresponding author: E-mail: hausinge@msu.edu

Contents

1. Introduction to CsiD 64
2. Functional assignment of CsiD 67
   2.1 Structural assignment of CsiD as an Fe(II)/2OG-dependent oxygenase 67
   2.2 Biochemical confirmation that CsiD is an Fe(II)/2OG oxygenase 70
   2.3 Identification of CsiD as a glutarate L-2 hydroxylase 71
3. Characterization of glutarate L-2-hydroxylase 73
   3.1 Enzyme kinetics of glutarate L-2-hydroxylase 73
   3.2 Structures of glutarate L-2-hydroxylase:ligand complexes 75
   3.3 Reaction mechanism of glutarate L-2-hydroxylase 78
4. Distribution of glutarate L-2-hydroxylase 80
5. Cellular role of glutarate L-2-hydroxylase 81
6. Conclusions 84
7. Experimental section 84
   7.1 Strain construction 84
   7.2 Purification of CsiD 85
   7.3 Product analyses and enzyme assays 86
Acknowledgments 86
References 86

[1] Present address: Department of Biology, Central College, Pella, IA, United States.
[2] Present address: Regulatory Affairs Department, Lonza, Slough, Berkshire, United Kingdom.

Advances in Protein Chemistry and Structural Biology, Volume 117
ISSN 1876-1623
https://doi.org/10.1016/bs.apcsb.2019.05.001
© 2019 Elsevier Inc.
All rights reserved.

Abstract

The *Escherichia coli* gene initially named *ygaT* is located adjacent to *lhgO*, encoding L-2-hydroxyglutarate oxidase/dehydrogenase, and the *gabDTP* gene cluster, utilized for γ-aminobutyric acid (GABA) metabolism. Because this gene is transcribed specifically during periods of carbon starvation, it was renamed *csiD* for carbon starvation induced. The CsiD protein was structurally characterized and shown to possess a double-stranded ß-helix fold, characteristic of a large family of non-heme Fe(II)- and 2-oxoglutarate (2OG)-dependent oxygenases. Consistent with a role in producing the substrate for LhgO, CsiD was shown to be a glutarate L-2-hydroxylase. We review the kinetic and structural properties of glutarate L-2-hydroxylase from *E. coli* and other species, and we propose a catalytic mechanism for this archetype 2OG-dependent hydroxylase. Glutarate can be derived from L-lysine within the cell, with the *gabDT* genes exhibiting expanded reactivities beyond those known for GABA metabolism. The complete CsiD-containing pathway provides a means for the cell to obtain energy from the metabolism of L-lysine during periods of carbon starvation. To reflect the role of this protein in the cell, a renaming of *csiD* to *glaH* has been proposed.

1. Introduction to CsiD

In 1993, a series of carbon starvation-induced genes was identified in *Escherichia coli* on the basis of induction studies using *lacZ* fusions created with a λp*lac*Mu system (Weichart, Lange, Henneberg, & Hengge-Aronis, 1993). One of these genes, initially termed *csi-12* or *ygaT* and later designated *csiD*, was shown to require the general stress sigma factor RpoS (σ^s) (Hengge, 2008), 3′,5′-cyclic adenosine monophosphate (cAMP), and the cAMP-receptor protein (CRP) for its transcription (Marschall et al., 1998). Subsequent investigation (Metzner, Germer, & Hengge, 2004) revealed that this gene initiates a six-gene cluster: *csiD*, *ygaF* (later renamed *lhgO*), *gabD*, *gabT*, *gabP*, and *csiR* (Fig. 1). Regulation of these genes is complex and includes action by the CsiR repressor (Metzner et al., 2004) and at least three promoters: $csiD_p$ is used during carbon starvation or stationary phase and dependent on cAMP, CRP, the leucine regulatory protein (Lrp), and σ^s; $gabD_{p1}$ is governed by σ^s; and $gabD_{p2}$ is controlled by σ^s and Nac (Marschall et al., 1998; Metzner et al., 2004; Schneider et al., 2002).

Biochemical roles have been identified for the *lhgO* and *gabDTP* gene products (Fig. 2). The name *lhgO* derives from its function; it encodes L-2-hydroxyglutarate (L-2HG) oxidase, a soluble flavoprotein that reduces dioxygen to hydrogen peroxide as it converts L-2HG to 2-oxoglutarate

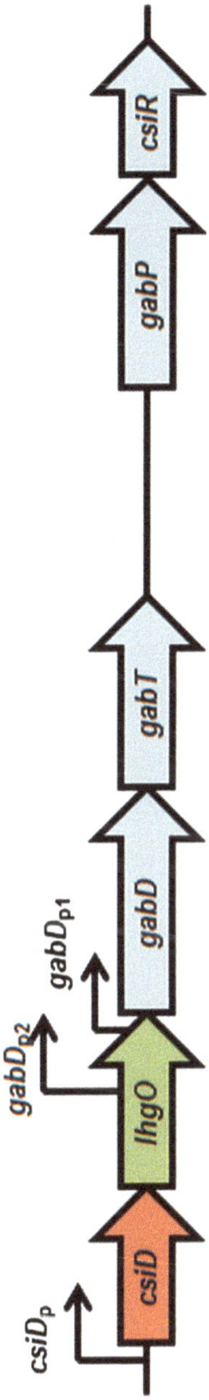

Fig. 1 Genomic location of *csiD* in *E. coli* and regulation of the *csiD-lhgO-gabDTP-csiR* gene cluster.

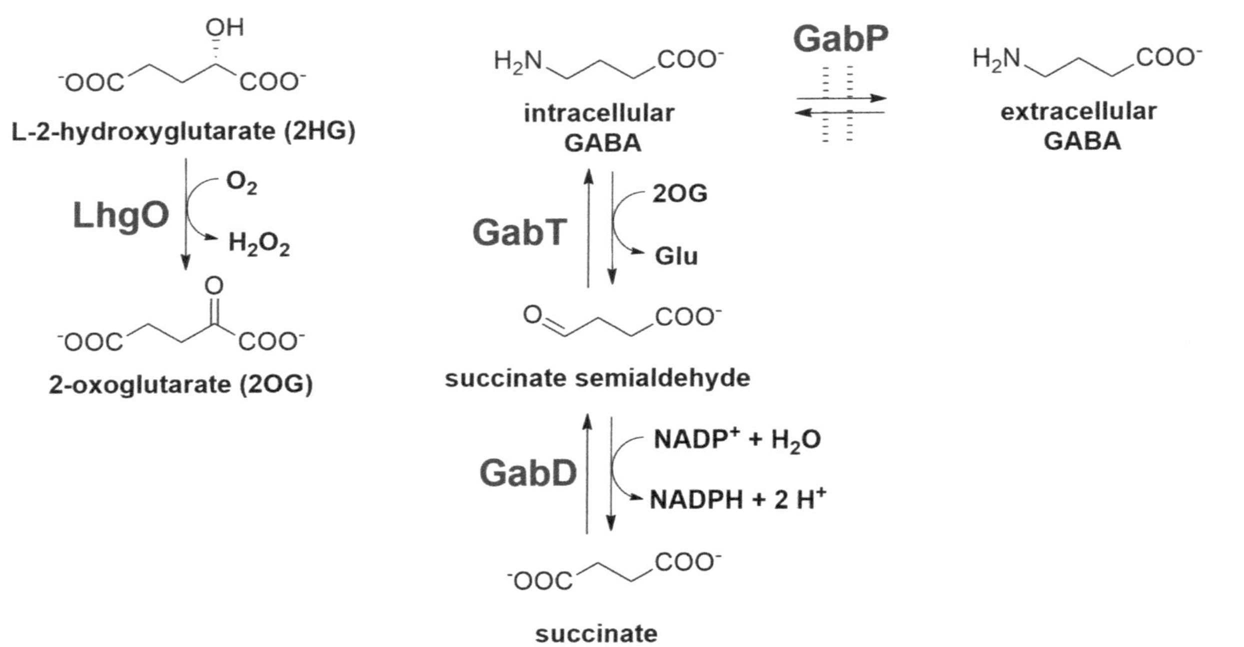

Fig. 2 Functions of LhgO and GabDTP. LhgO catalyzes the O_2-dependent oxidation (as illustrated) or the menaquinone-linked dehydrogenation (not shown) of L-2HG to form hydrogen peroxide (or reduced menaquinone) and 2OG. The GabP permease allows GABA to traverse the cell membrane (indicated by the *stippled lines*). The GabT transaminase converts intracellular GABA into succinic acid semialdehyde while transforming 2OG into glutamic acid (Glu). GabD then catalyzes the $NADP^+$-dependent oxidation of succinic acid semialdehyde to generate succinate.

(2OG) (Kalliri & Hausinger, 2008). A soluble L-2HG oxidase activity also was demonstrated for the corresponding LhgO from *Pseudomonas putida* KT2440 (Zhang et al., 2018). Additional studies of the *E. coli* protein found it to be a membrane-bound, ubiquinone-linked dehydrogenase (Knorr et al., 2018). It is unclear why the two studies of *E. coli* LhgO report differences in cellular localization and function, but the discrepancies may relate to subtle differences associated with the pET42b versus pET16b plasmid constructs. The *gabDTP* cluster encodes an $NADP^+$-dependent succinic acid semialdehyde dehydrogenase (Cozzani, Fazio, Felici, & Barletta, 1980), a pyridoxal phosphate-dependent γ-aminobutyrate (GABA) transaminase (Dover & Halpern, 1974), and a GABA-specific membrane-bound permease (Niegemann, Schulz, & Bartsch, 1993), together allowing the cells to metabolize extracellular GABA to generate intracellular succinic acid. These studies did not address the cellular role of CsiD.

2. Functional assignment of CsiD

2.1 Structural assignment of CsiD as an Fe(II)/2OG-dependent oxygenase

Crystal structures were reported for CsiD from *E. coli* strain K-12 that was truncated by 14-residues at its *N*-terminus and from *E. coli* strain B that was full-length, but with the first 14 residues disordered (Table 1; PDB ID: 1JR7and 2R6S, respectively) (Chance et al., 2002; Lohkamp & Dobritzsch, 2008). The sequences of these proteins differ at six positions (Ala68/Gly68, Val102/Ile102, Lys106/Ala106, Asn201/His201, Ser318/Thr318, and Asn319/His319 for strains K-12/B, respectively), possibly accounting for the observed tetrameric versus octameric crystal structures (although differences in pH alternatively may have impacted the quaternary structure). In solution, both proteins behave as tetramers of a subunit that possesses a double-stranded ß-helix fold containing a ferrous ion bound to a 2-histidine-1-carboxylate motif involving His160, Asp162, and His292 (Fig. 3). This protein fold and metallocenter arrangement is a hallmark of Fe(II)/2OG-dependent oxygenases (Aik et al., 2015), a family of enzymes that catalyzes hydroxylation, halogenation, ring formation, desaturation, epimerization, and other reactions (Hausinger, 2004; Herr & Hausinger, 2018; Martinez & Hausinger, 2015). The diversity of potential reactions and the known breadth of substrates utilized by Fe(II)/2OG-dependent oxygenase family members, including proteins, oligonucleotides, lipids, amino acids, bases,

Table 1 Summary of CsiD crystal structures.

PDB ID	Metal	Ligand(s)	pH and precipitates	Space group	Cell constants a, b, c (Å) α, β, γ (°)	Resolution (Å)
1JR7	Fe	None	pH 5.6; $(NH_4)_2SO_4$	I 4 2 2	120, 120, 136 90, 90, 90	2.0
2R6S	Fe	None	pH 9, 1.6 M $(NH_4)_2SO_4$	I 4 2 2	120, 120, 137 90, 90, 90	2.1
6GPE	Fe	None	pH 6, 30% 2-methylpentanediol	P 4 2_1 2	121, 121, 137 90, 90, 90	2.2
6HL8	Fe	Glutarate	pH 6, 30% 2-methylpentanediol	P 4 2_1 2	120, 120, 136 90, 90, 90	2.4
6GPN	Fe	NOG[a]	pH 7, 3.5 M sodium formate	P 4 2_1 2	121. 121, 137 90, 90, 90	2.2
6HL9	Fe	Succinate	pH 7, 1 M succinate	P 4 2_1 2	123, 123, 138 90, 90, 90	2.3

[a]*NOG*, N-oxalylglycine.

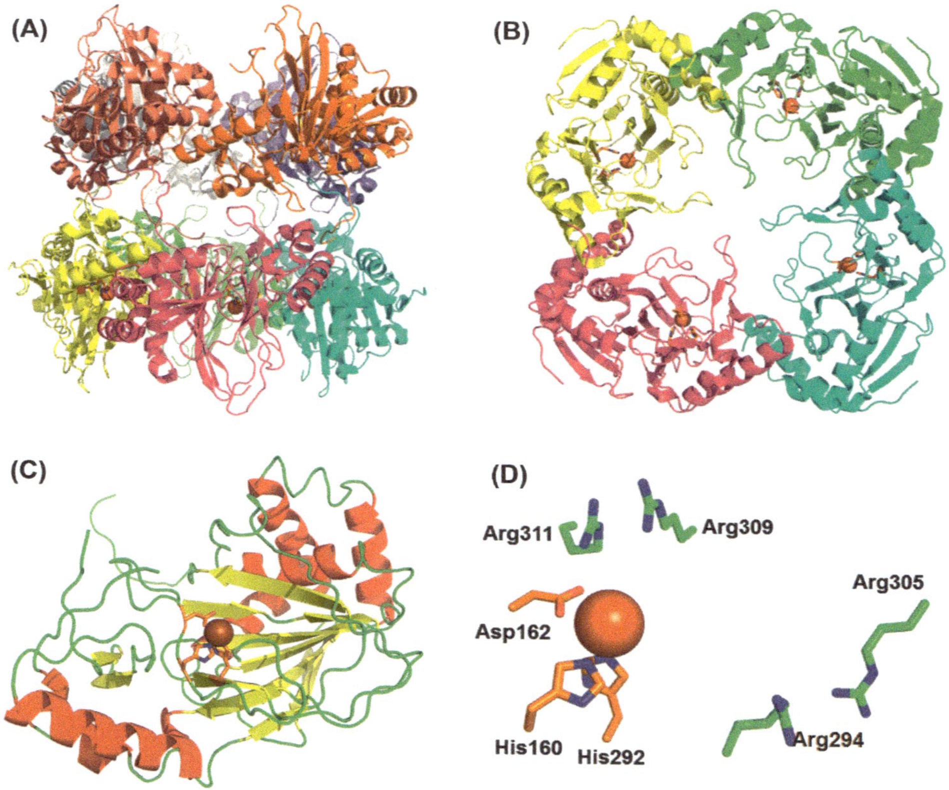

Fig. 3 Overall structure of CsiD. The CsiD (A) octamer, (B) tetramer, and (C) monomer are depicted in cartoon mode with different subunits in the oligomers shown in distinct colors and the single subunit shown with ß-strands in yellow, α-helices in red, and loops in green. Fe atoms are shown as brown spheres and the protein ligands to the metal are shown as orange sticks. (D) An expanded view of the active site depicts several positively-charged arginine side chains near the active site that could interact with 2OG and the primary substrate.

terpenoids, and more (Hausinger, 2015; Herr & Hausinger, 2018; Islam, Leissing, Chowdhury, Hopkinson, & Schofield, 2018; Wu, Meng, & Tang, 2016), initially hindered assignment of the primary substrate of CsiD and its cellular function.

2.2 Biochemical confirmation that CsiD is an Fe(II)/2OG oxygenase

Because of our long-standing interest in Fe(II)/2OG-dependent oxygenases (Hausinger, 2004, 2015; Herr & Hausinger, 2018; Martinez & Hausinger, 2015) and as an extension of our earlier characterization of LhgO from *E. coli* (Kalliri & Hausinger, 2008), we initiated investigations of CsiD from this microorganism. We found that CsiD catalyzes an uncoupled reaction in which oxygen and 2OG are consumed in the absence of the enzyme's primary substrate if an external reductant such as ascorbic acid is provided, as previously noted for several other members of the Fe(II)/2OG dioxygenase family (Hausinger, 2004). For example, incubation of CsiD (2 μM final concentration) in 25 mM imidazole buffer, pH 6.75, containing 100 μM ferrous ions, 200 μM ascorbate, and 1 mM 2OG led to oxygen consumption at an initial rate of 700–900 nmol min^{-1} (mg $protein)^{-1}$ when measured with a Clark-type oxygen electrode. Furthermore, a similar enzyme preparation (2.5 μM CsiD in 25 mM imidazole buffer, pH 6.75, containing 50 μM ferrous ions, 0.4 mM ascorbate, and 1 mM 2OG) resulted in 2OG consumption and succinate production at initial rates of 738 nmol min^{-1} (mg $protein)^{-1}$ and 612 nmol min^{-1} (mg $protein)^{-1}$, respectively, when measured by HPLC. These results support the assignment of CsiD as a 2OG-dependent oxygenase.

Another diagnostic property of members within the Fe(II)/2OG-dependent oxygenase family is the development of a pink or purple color when anaerobic apoprotein is incubated with Fe(II) and 2OG (Proshlyakov, McCracken, & Hausinger, 2017). This chromophore has been attributed to metal-to-ligand charge-transfer (MLCT) transitions associated with the bidentate coordination of the metal by the C1 carboxylate and C2 keto group of 2OG (Pavel et al., 1998). CsiD was treated with EDTA to remove any contaminating nickel or other metal ions that might be bound to the purified protein. After dialysis against EDTA-free buffer containing 25 mM Tris and 1 mM dithiothreitol, an Fe(II)/ascorbate mixture was added, followed by addition of 2OG to yield the difference spectrum in Fig. 4. The position and intensity of the λ_{max} (510 nm with $\varepsilon_{510} = 158\ M^{-1}\ cm^{-1}$) is similar to the situation for many other proteins of this family such as the sulfonate-degrading enzyme TauD, the

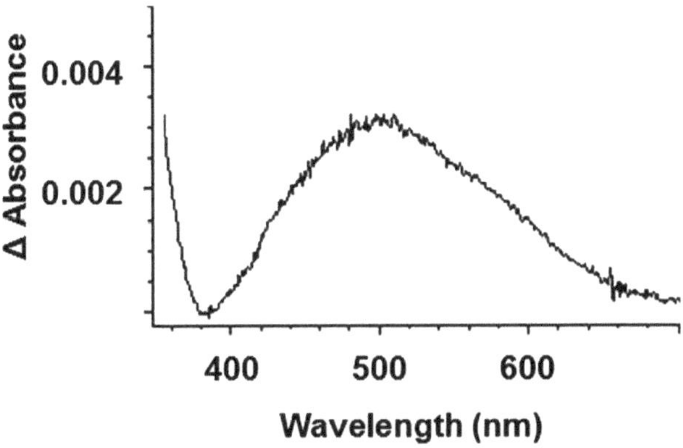

Fig. 4 CsiD exhibits a visible absorbance spectrum in the presence of Fe(II) and 2OG. Anaerobic solutions of CsiD were prepared in sealed serum vials by alternative cycles of vacuum treatment and argon purging. Stock solutions of 2OG (100 mM) were prepared in 100 mM HEPES buffer, pH 8.0, and made anaerobic by 5 rounds of degassing and flushing with argon. Mixtures of ferrous ammonium sulfate (18 mM) and sodium dithionite (100 mM) were prepared by three rounds of degassing and argon flushing of the solids, followed by addition of the desired volume of anoxic water. The protein (25 μM) was mixed with 125 μM $(NH_4)_2Fe(SO_4)_2$ and 1 mM 2OG in 25 mM Tris buffer, pH 8.0, at room temperature and the difference spectrum was obtained after subtracting the spectrum of the same concentrations of enzyme and iron.

herbicide-metabolizing representative TfdA, and the biosynthetic enzyme clavaminate synthase (λ_{max} 500–530 nm and ε_{530} in the range of 140–375 M^{-1} cm^{-1}) (Hegg et al., 1999; Pavel et al., 1998; Ryle, Padmakumar, & Hausinger, 1999). These results confirm that CsiD binds ferrous ion plus 2OG in a bidentate coordination complex.

2.3 Identification of CsiD as a glutarate L-2 hydroxylase

Because *E. coli csiD* lies adjacent to *lhgO* encoding an L-2HG oxidase (Kalliri & Hausinger, 2008) or L-2HG dehydrogenase (Knorr et al., 2018), we reasoned that CsiD may be a 2OG-dependent glutarate L-2-hydroxylase. To test this hypothesis, we incubated CsiD with equal amounts of 2OG and glutaric acid while monitoring the substrate mixture and products by ^1H NMR spectroscopy. We observed the complete consumption of both 2OG and glutarate with conversion of the substrates to 2HG and succinate (Fig. 5). Essentially identical NMR results were published by others (Knorr et al., 2018). Confirmation of this reaction was obtained by mass spectrometric analyses with the *E. coli* protein (Knorr et al., 2018) and by HPLC quantifications using an organic acids column with the *P. putida* enzyme (Zhang et al., 2018).

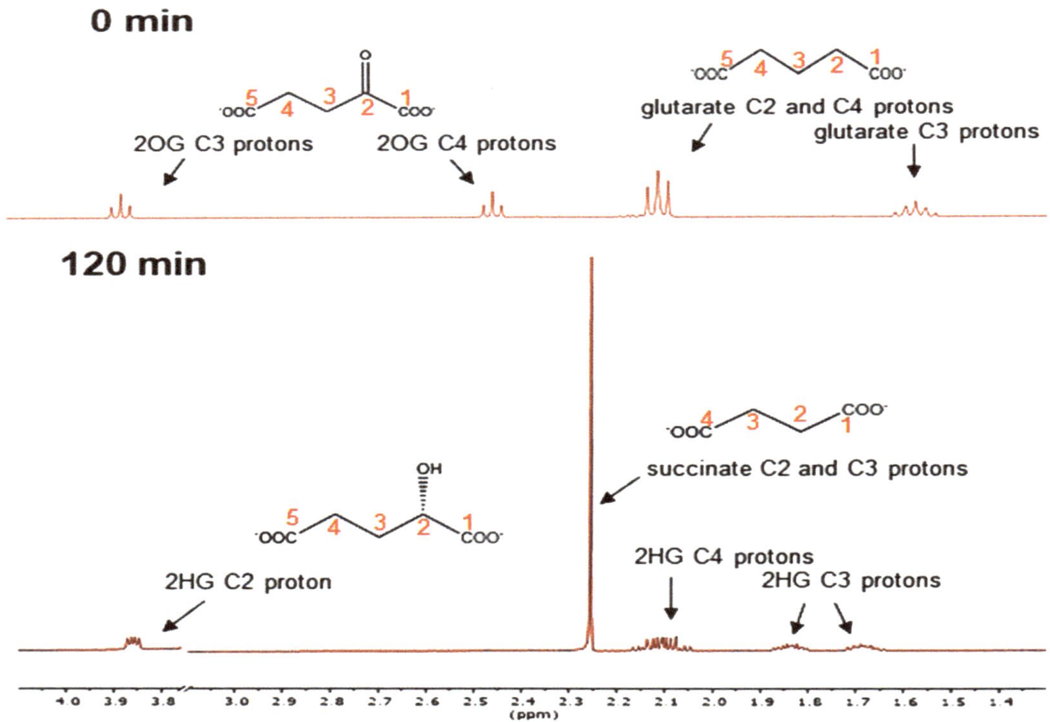

Fig. 5 CsiD converts 2OG and glutarate into succinate and 2HG as shown by ^{1}H NMR spectroscopy. CsiD (2 μM) was incubated with 1 mM 2OG, 1 mM glutarate, and 0.1 mM ferrous ions in 0.7 mL of 1 mM Tris buffer (pH 8.0) at 37 °C. Samples were boiled for 3 min at 100 °C and centrifuged at 10,000 *g* for 10 min, then the supernatants were subjected to vacuum in a Speed-Vac until the solvent had evaporated. The remaining solids were dissolved in deuterated water and 300 MHz proton ^{1}H NMR spectra were obtained. The upper panel depicts the spectrum obtained at zero time with peaks corresponding to 2OG and glutarate. The lower panel shows the spectrum after 2 h of incubation revealing the disappearance of substrates and formation of succinate and 2HG.

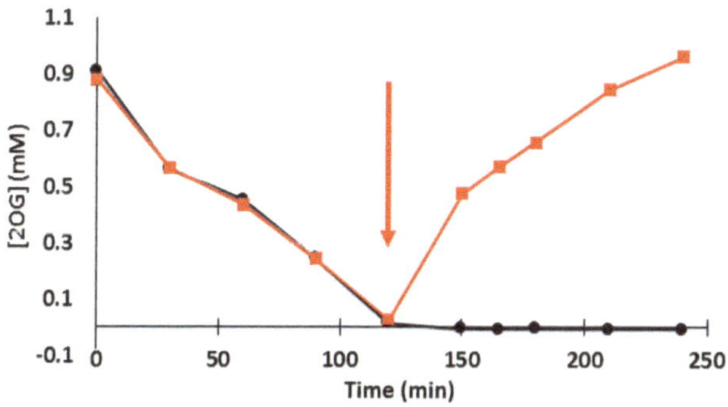

Fig. 6 CsiD generates the L-isomer of 2HG. CsiD (2 μM) was incubated with 1 mM glutarate, 1 mM 2OG, and 0.1 mM ferrous ions in 100 mM HEPES buffer (pH 7.0). Aliquots (250 μL) were removed at the times indicated and the OPD assay was used to monitor the remaining 2OG. LhgO (2 μM, red squares) was added to the sample at the arrow and the recovery of 2OG was assessed using the OPD assay.

To assess the enantiospecificity of the 2HG produced by CsiD, we examined the effect of mixing the product of the CsiD reaction with LhgO that is known to be specific for L-2HG (Kalliri & Hausinger, 2008). A CsiD reaction mixture was incubated at 37 °C and timed aliquots were examined for the remaining concentrations of 2OG by colorimetric quantification using *ortho*-phenylenediamine (OPD), a reagent that reacts with ketones and aldehydes (Hayashi, Tsuchiya, Todoriki, & Naruse, 1982). After the 2OG was consumed at 2 h, LhgO was added to convert L-2HG to 2OG that was quantified by the OPD assay. The levels of 2OG generated were stoichiometric with the amount originally provided (Fig. 6), thus demonstrating that CsiD specifically forms the L-isomer of 2HG. Formation of the L-isomer product also was demonstrated by derivatization with diacetyl-L-tartaric anhydride and analysis by liquid chromatography-mass spectrometry (Knorr et al., 2018). In addition, L-specificity was inferred for the product obtained by *P. putida* CsiD on the basis of its inability to be utilized by a D-specific 2HG dehydrogenase (Zhang et al., 2018).

3. Characterization of glutarate L-2-hydroxylase

3.1 Enzyme kinetics of glutarate L-2-hydroxylase

Selected kinetic parameters have been obtained for CsiD isolated from *E. coli*, *P. putida*, *Klebsiella pneumoniae*, and *Salmonella typhimurium* (Knorr et al., 2018; Zhang et al., 2018) (Table 2). The two specific activity values

Table 2 Kinetic parameters of CsiD from different sources.

Source of CsiD	Specific activity (μmol min^{-1} mg^{-1})	Glutarate K_m (μM)	2OG K_m (μM)	References
E. coli	53 ± 3	650 ± 20	100 ± 7	Knorr et al., 2018
E. coli	126.8 ± 0.4	NR[a]	NR	Zhang et al., 2018
P. putida	177 ± 3	145.7 ± 1.5	95.3 ± 1.0	Zhang et al., 2018
K. pneumoniae	118.1 ± 0.4	NR	NR	Zhang et al., 2018
S. typhimurium	194.2 ± 0.3	NR	NR	Zhang et al., 2018

[a]*NR*, not reported.

reported in studies of the *E. coli* protein were more than twofold different, perhaps due to the distinct assay conditions used. Although both measurements utilized an oxygen electrode and included 400 μM ascorbate and 1 mM 2OG, one was carried out in 100 mM MOPS buffer (pH 7.2) containing 5 mM glutarate, 70 mM KCl, and 4 μM $Fe(NH_4)_2SO_4$ at 30 °C (Knorr et al., 2018), whereas the other was performed in 20 mM imidazole (pH 6.7) containing 1 mM glutarate, no added salt, and 50 μM ferrous ions at an unstated temperature (Zhang et al., 2018). The specific activity values for CsiD from all sources are the same order of magnitude, and for *E. coli* the glutarate 2-hydroxylase activity corresponds to at least 100-fold the uncoupled rate for oxidative decarboxylation of 2OG in the presence of ascorbate (Section 2.2). The K_m parameters were determined only for the *E. coli* and *P. putida* proteins; values for glutarate were more than fourfold different, whereas those for 2OG were identical. In the case of *P. putida* CsiD, the K_m of oxygen (266 ± 14 μM) also was determined (Zhang et al., 2018). Although k_{cat} and k_{cat}/K_m values for glutarate, 2OG, and oxygen were provided in the *P. putida* CsiD study, these numbers are inconsistent with the reported specific activity and V_{max} values (Zhang et al., 2018) and should not be used when comparing to such values for enzymes from other sources.

CsiD appears to be highly specific for glutaric acid as the primary substrate, with no activity observed when this compound was substituted by 5-aminovalerate, 2-aminoadipate, 2-oxoadipate (2OA), succinate, adipate, L-lysine, or glutaryl-CoA using the *P. putida* enzyme (Zhang et al., 2018). Studies with the *E. coli* CsiD found no activity when substituting glutarate with other dicarboxylic acids including oxalate, malonate, succinate, adipate, and pimelate. Furthermore, CsiD proteins from *Halobacillus* sp. BAB-2008, *P. putida*, and *E. coli* were confirmed to lack activity using malonate, succinate, adipate, propionate, butyrate, pentanoate, and hexanoate while retaining the capacity to hydroxylate glutarate (Thompson et al., 2019). While 2OA was not used as the primary substrate, *E. coli* CsiD is active when 2OA replaces 2OG in the reaction (unpublished observations and Thompson et al., 2019). Several other family members can substitute 2OG with 2OA, a 2-oxo acid containing an extra methylene carbon, for catalyzing their reactions (e.g., Eichhorn, van der Ploeg, Kertesz, & Leisinger, 1997; Fukumori & Hausinger, 1993; Kahnert & Kertesz, 2000; Martinez & Hausinger, 2016).

Inhibitors of CsiD include a variety of metal chelators, which remove the essential ferrous ion from the active site (unpublished observations), and *N*-oxalylglycine (NOG), a 2OG mimic that is incapable of supporting catalysis (Knorr et al., 2018). These compounds are known to inhibit a multitude of other Fe(II)/2OG-dependent oxygenase (Aik et al., 2015).

3.2 Structures of glutarate L-2-hydroxylase:ligand complexes

Knorr et al. describe four crystal structures of CsiD (Fig. 7) that provide keen insights into how the protein interacts with it substrates (Knorr et al., 2018). In agreement with the previously described structures of this protein (Chance et al., 2002; Lohkamp & Dobritzsch, 2008), the catalytic metal ion is bound within the double-stranded ß-helix fold by the His160, Asp162, and His292 sidechains, with the octahedral coordination geometry completed by three water molecules (Fig. 7A). Several positively-charged Arg residues located in the active site pocket are potentially poised to interact with the carboxylate-containing substrates (2OG and glutarate).

The structure of the 2OG-bound enzyme was approximated by using the 2OG analog NOG (Fig. 7B). Structures have been solved for many other Fe(II)/2OG oxygenase family members in the presence of NOG or 2OG, and they typically exhibit bidentate metal coordination of the ligand with the C2 carbonyl group opposite the carboxylate side chain and the C1 carboxyl group located opposite one of the His ligands or at an intermediate position. Furthermore, most other family representatives possess two arginine residues at the active site; one Arg forms a hydrogen bond with the proximal carboxylate of 2OG/NOG and a second Arg stabilizes the distal carboxylate of 2OG/NOG in a salt bridge (Aik et al., 2015). For CsiD, NOG coordinates the metal in a bidentate manner with its C2 carbonyl located opposite Asp162 and the C1 carboxylate positioned approximately mid-way opposite the two His residues. A hydrogen bond from Arg309

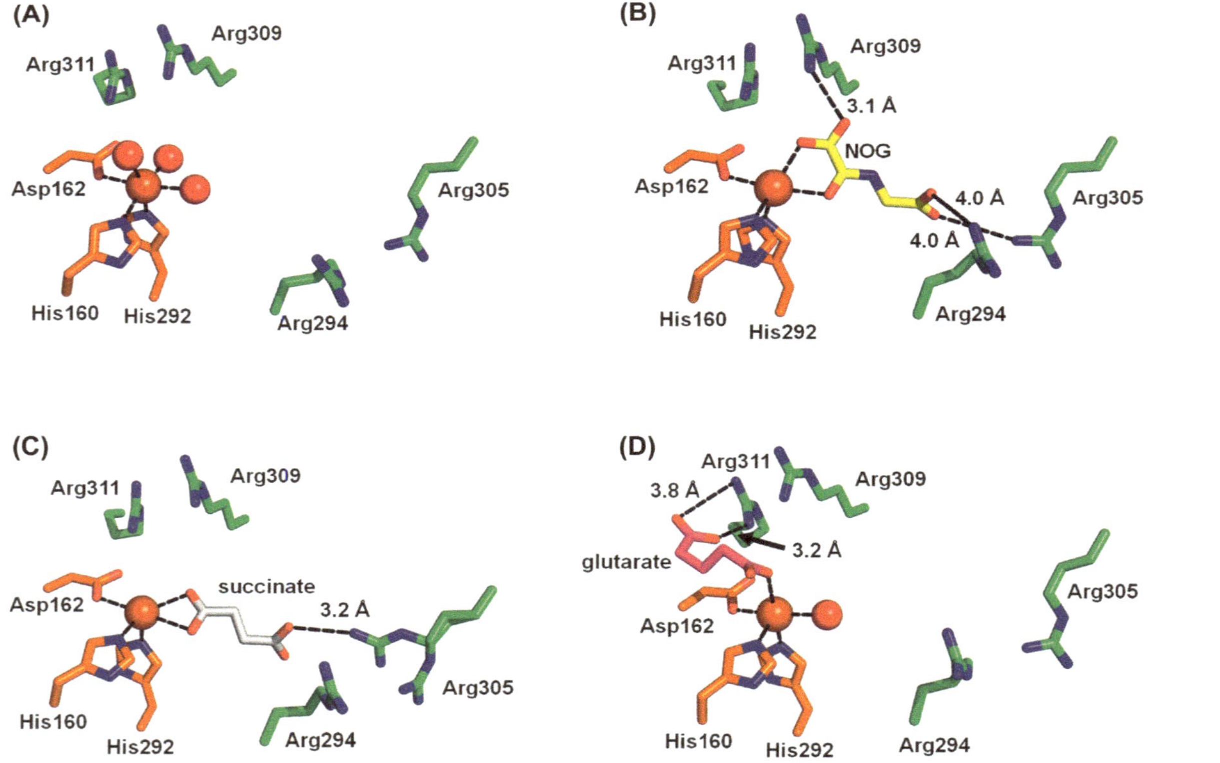

Fig. 7 Ligand-bound structures of glutarate L-2-hydroxylase: (A) CsiD holoprotein (PDB code: 6GPE), (B) CsiD:NOG (PDB code: 6GPN), (C) CsiD:succinate (PDB code: 6HL9), and (D) CsiD:glutarate (PDB code: 6HL8). Fe(II) atoms are shown as brown spheres, coordinated waters are red spheres, and metal-binding residues are shown as sticks with orange carbon atoms. Selected Arg side chains (green carbon atoms) near the active site are depicted, with panel C showing two conformations for Arg305. Selected distances are indicated.

stabilizes the metal-bound carboxyl group, and very long hydrogen bonds are potentially provided by Arg294 and Arg305 to the distal carboxylate.

The structure of CsiD with the bound product succinate (Fig. 7C) is notable in two respects. First, the metal coordination geometry involving His160, Asp162, and the two carboxylate oxygen atoms of the product is nearly planar, leading to an open coordination site opposite His292. This open site indicates the most likely position for the L-2HG hydroxyl oxygen atom if that product were present. Second, the distal carboxylate of succinate is hydrogen bonded to one of two conformations of Arg305 (whereas the distance of this carboxylate of succinate to Arg294 has expanded to 5.2 Å). The flexibility of the Arg305 side chain suggests that it also may be capable of adopting a conformation that could form a salt bridge with this product or with NOG/2OG.

The glutarate-bound structure (Fig. 7D) depicts a severely bent substrate with one carboxylate forming a long and asymmetric salt bridge (3.2 and 3.8 Å in length) with a highly twisted Arg311 side chain (and the backbone amide of Gly163, not shown). The second carboxyl group of glutarate binds to the metal ion from which two water molecules have been displaced. This aspect of the binding mode is unlikely to be relevant to the situation that occurs when both glutarate and 2OG are coordinated to the protein during catalysis. In particular, 2OG is nearly certain to occupy two metal coordination sites and one site must be available to bind oxygen, so the glutarate-Fe(II) bond is not catalytically significant. Furthermore, the C2 carbon atoms of glutarate are inappropriately positioned for hydride abstraction by an Fe(IV)-oxo intermediate in order to yield the L-2HG product.

We generated a hypothetical model of the 2OG- and glutarate-bound structure of CsiD (Fig. 8). This view was created from the structure of NOG-bound CsiD by substituting the N atom of NOG with a carbon atom and rotating the position of the C1 carboxylate to lie closer to the plane formed by His160, Asp162, and the Fe(II) atom. A salt bridge between the 2OG C5 carboxylate and Arg305 was formed by rotating bonds of the arginine side chain. Small bond rotations in Arg309 allowed for hydrogen bonding to the 2OG C1 carboxylate and to glutaric acid. The substrate glutarate was shifted in position from that found in the glutarate-only structure and made less bent. Furthermore, bond rotations of Arg311 and Arg294 allowed for salt bridge stabilization of each carboxylate. This pose of the substrate is notable because the methylene carbon adjacent to the Arg311-interacting carboxylate is well positioned for hydrogen abstraction followed by hydroxyl radical rebound to yield the correct enantiomer:

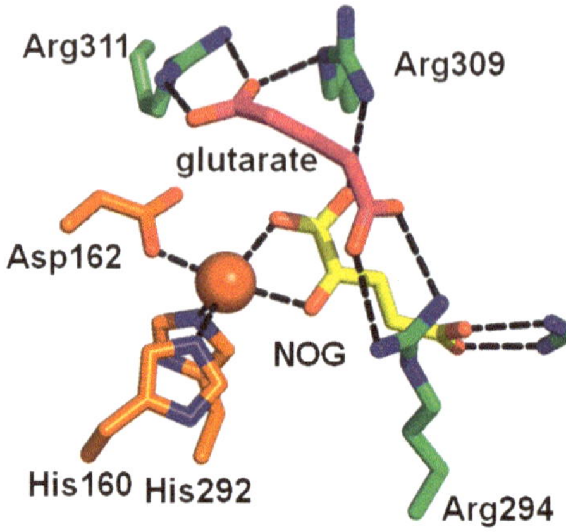

Fig. 8 Hypothetical view of the CsiD active site with bound 2OG (*yellow*) and glutarate (*red*).

L-2HG. We stress that this figure presents only a theoretical view and a crystal structure of the authentic complex is needed for verification.

3.3 Reaction mechanism of glutarate L-2-hydroxylase

On the basis of the known properties of glutarate L-2-hydroxylase described above along with the wealth of information available for other family members (Schofield & Hausinger, 2015), we propose the enzymatic reaction mechanism shown in Fig. 9. The resting state of the enzyme (boxed) possesses Fe(II) bound to three protein side chains (His160, Asp162, and His292) and with three coordinating water molecules. Bidentate binding of 2OG via a C1 carboxylate oxygen (opposite His160) and C2 keto oxygen (opposite Asp262) results in the release of two water molecules, generating the MLCT transitions that are diagnostic of this enzyme family (Fig. 4). The binding of 2OG is stabilized by Arg305 forming a salt bridge with the C5 carboxylate and by Arg309 hydrogen bonding to the C1 carboxylate (not depicted in the figure). The primary substrate, glutarate, binds near the metallocenter, but without direct coordination to the metal and possibly forms salt bridges with Arg311 and Arg294 (Fig. 8). The binding of the primary substrate results in perturbation of the MLCT features (data not shown), displacement of the final water molecule, and creation of an oxygen binding site. Binding of oxygen likely results in an Fe(III)-superoxo state that has not been directly observed in any representative of this group of enzymes. The reaction scheme next shows a plausible series of sequential steps involving a cyclic peroxo-Fe(IV) species (also not directly observed for any family

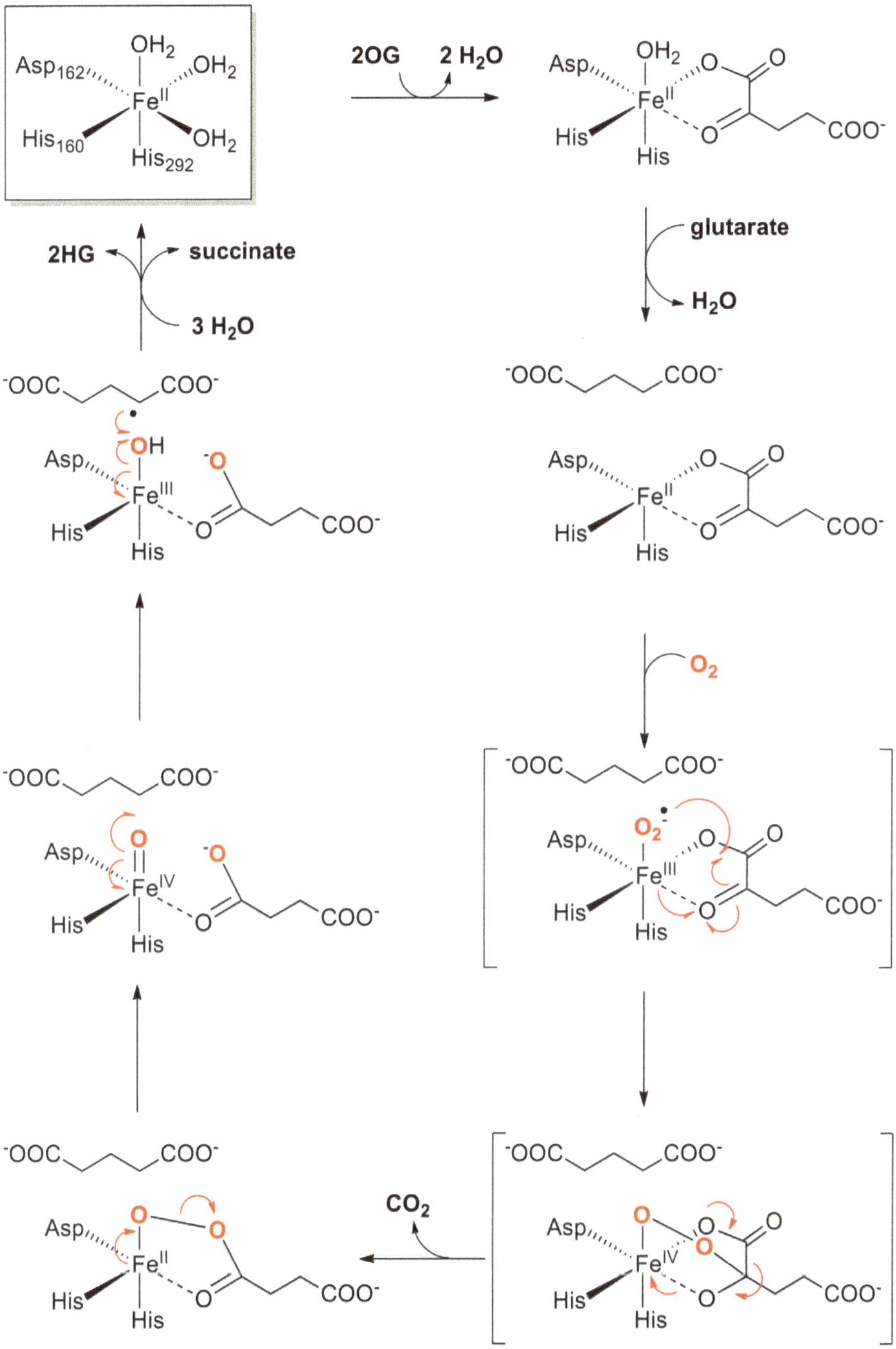

Fig. 9 Proposed enzymatic reaction mechanism of glutarate L-2-hydroxylase. Potential electron transfer reactions are shown by red arrows and the oxygen atoms derived from dioxygen are indicated in bold red. Two species that have not yet been directly detected in any 2OG-dependent oxygenase are shown in brackets.

member), a persuccinate bound Fe(II) (crystallographically observed in VioC; Mitchell et al., 2017), and an Fe(IV)-oxo (ferryl) species; the pathway to reach the ferryl species remains uncertain, but direct evidence for the existence of the ferryl state has been obtained for several other members of this enzyme family (Chang et al., 2014; Chang, Li, Lee, Cronican, & Guo, 2016; Dunham et al., 2018; Galonić, Barr, Walsh, Bollinger, & Krebs, 2007; Hoffart, Barr, Guyer, Bollinger, & Krebs, 2006; Matthews et al., 2009; Mitchell et al., 2017; Price, Barr, Glass, Krebs, & Bollinger, 2003; Price, Barr, Tirupati, Bollinger, & Krebs, 2003; Proshlyakov, Henshaw, Monterosso, Ryle, & Hausinger, 2004). Accompanying formation of the ferryl species is the production of succinate and carbon dioxide.

The ferryl intermediate is positioned appropriately for in-line (Hausinger, 2004) abstraction of the hydrogen atom from the L prochiral position of the glutarate C2 atom, yielding Fe(III)-hydroxide and a substrate radical. Hydroxyl radical rebound would then yield L-2HG and restore the Fe(II) state of the enzyme. Alternatively, deprotonation of Fe(III)-hydroxide may lead to an Fe(III)-oxo state that reacts with the substrate radical to produce glutarate alkoxide bound to Fe(II) (not shown in Fig. 9), as suggested for other enzymes in this family on the basis of spectroscopy (Grzyska, Appelman, Hausinger, & Proshlyakov, 2010), crystallographic evidence (Mitchell et al., 2017), or computational methods (Fang & Cisneros, 2013). In either case, product dissociation and rebinding of three water molecules completes the cycle.

4. Distribution of glutarate L-2-hydroxylase

Zhang et al. identified homologs of *csiD* in 454 *Proteobacteria* and homologs of *lhgO* in 608 *Proteobacteria* and 4 *Actinobacteria* (Zhang et al., 2018). The distribution of each gene was suggested to be narrow and sporadic. Whereas genes encoding CsiD appear to be produced only in bacteria, those encoding LhgO also occur in eukaryotic microbes, animals, and plants. While 164 genomes included only *lhgO* (perhaps used for synthesis of a repair enzyme to remove this harmful metabolite), 448 genomes had both *csiD* and *lhgO*, and only 6 genomes had *csiD* without *lhgO*. The phylogenetic distribution of *csiD* also was assessed using the refseq database from May 2018 (Knorr et al., 2018); homologs are found in α-, β- γ-, and ε-proteobacteria and bacilli.

5. Cellular role of glutarate L-2-hydroxylase

Deletion mutants of *csiD* (Δ*csiD*) exhibit significant growth phenotypes compared to wild-type cells. For example, *E. coli* BL21(DE3) cells in the late stages of growth benefited by the inclusion of glutarate in minimal glucose medium, whereas a corresponding Δ*csiD* culture reached stationary phase regardless of the presence of this supplemental nutrient (Fig. 10). Furthermore, the Δ*csiD* strain of *E. coli* BW25113 leads to a growth defect after reaching an OD_{600} of about 2 compared to that wild-type strain when grown in lysogeny broth (LB) medium (Knorr et al., 2018). In addition, deletion of *csiD* in *P. putida* results in delayed growth on glutarate medium (Zhang et al., 2018). These results are consistent with the notion that the Δ*csiD* strains fail to utilize glutaric acid during stationary phase growth and suggest that CsiD is utilized for metabolizing glutaric acid. In addition, a transposon insertion into *csiD* of *P. putida* results in decreased fitness when the cells are grown with 5-aminovalerate, an upstream metabolite in L-lysine degradation (Thompson et al., 2019).

Glutarate has been detected in mid-exponential phase *E. coli* and other cells growing on mineral salts medium with glucose provided as the carbon source (van der Werf, Overkamp, Muilwijk, Coulier, & Hankemeier, 2007). Furthermore, using gas chromatography-mass spectrometric analysis of cell extracts, we have shown that glutarate is present in stationary phase *E. coli* cells grown in TB medium (data not shown). When glutarate is generated in sufficient quantities within the *E. coli* cell, its metabolism via CsiD may be an important metabolic strategy—especially during stationary phase. In particular, the combined action of CsiD and LhgO allow cells to convert glutarate to CO_2 and succinic acid that can be further metabolized (Fig. 11).

The amino acid L-lysine appears to be the primary source of glutaric acid in both *E. coli* (Knorr et al., 2018) and *P. putida* (Zhang et al., 2018; Thompson et al., 2019). The *E. coli* pathway for lysine utilization is of special interest because of its ties to the GABA metabolism genes (Fig. 12). *E. coli* decarboxylates L-lysine to generate cadavarine by using two pyridoxal phosphate-dependent enzymes: LdcC and CadA (Lemonnier & Lane, 1998; Sabo, Boeker, Byers, Waron, & Fischer, 1974). The cadavarine is used to transaminate 2OG, forming glutamate and 5-aminopentanal, by the enzyme PatA, named for its analogous role in putrescine metabolism (Samsonova, Smirnov, Altman, & Ptitsyn, 2003). Similarly, PatD uses NAD^+ to oxidize 5-aminopentanal thus generating 5-aminovalerate in addition to its better

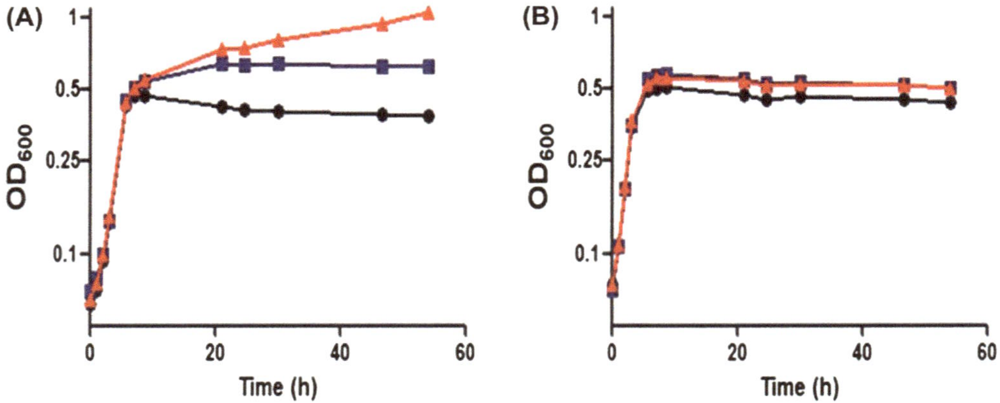

Fig. 10 CsiD increases cellular fitness in response to supplemental glutarate. (A) BL21(DE3) wild-type cells were grown in minimal medium A with 0.02% glucose and the presence of varying amount of glutarate (0 mM, circles; 1 mM, squares; and 5 mM triangles). (B) BL21(DE3) *csiD* cells were grown in identical conditions. The optical density at 600 nm (OD_{600}) was monitored to assess cell growth.

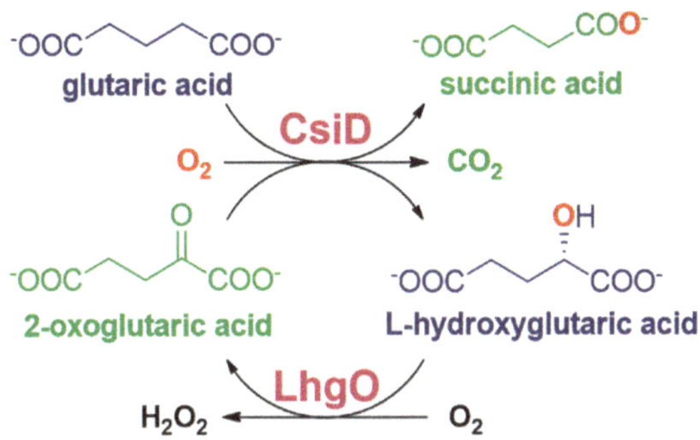

Fig. 11 CsiD and LhgO combine to allow the metabolism of glutaric acid to form succinic acid plus carbon dioxide.

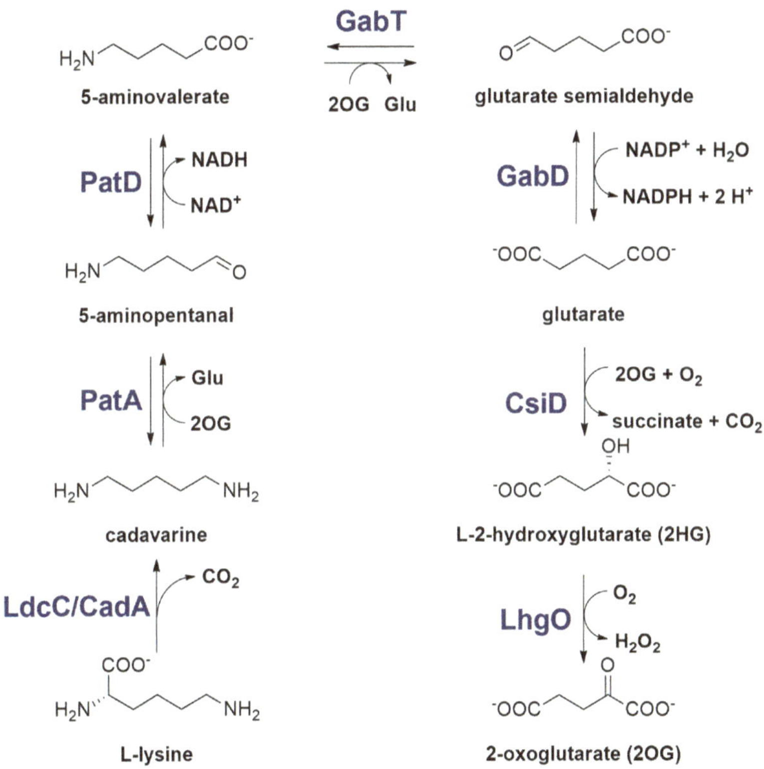

Fig. 12 Lysine degradation in stationary-phase *E. coli* cells occurs by a pathway that includes CsiD and LhgO. The functions shown for GabT and GabD are expanded beyond their well-known roles. The oxidase activity shown for LhgO may alternatively utilize a menaquinone-dependent dehydrogenase activity.

known function of converting 4-aminobutyraldehyde to GABA (Jorge, Perez-Garcia, & Wendisch, 2017; Knorr et al., 2018). Purified GabT and GabD were shown capable of converting 5-aminovalerate to glutarate with efficiencies comparable to those for conversion of GABA to succinate (Knorr et al., 2018). In *P. putida*, glutarate also is derived from L-lysine via 5-aminovalerate, but in this case the pathway uses the *davBA*, *davT*, and *davD* gene products (Liu et al., 2014; Revelles, Espinosa-Urgel, Molin, & Ramos, 2004; Zhang et al., 2018; Thompson et al., 2019). In both microorganisms, CsiD and LhgO complete the lysine degradation pathways. Glutarate also is available from several sources in the environment; this molecule is formed by various microbial species via hydrolysis of glutaryl-CoA, during benzoate degradation, via decomposition of L-tryptophan (or its metabolites including L-kynurenine, anthranilate, 2-aminomuconate, and 2-oxoadipate), and other pathways (Kanehisa et al., 2014).

6. Conclusions

CsiD is a 2OG-dependent glutarate L-2-hydroxylase that allows *E. coli* cells to obtain energy from L-lysine metabolism during stationary phase. Given this function, Knorr et al. suggest the corresponding gene name be replaced by *glaH* (for glutarate hydroxylase) (Knorr et al., 2018). The period of carbon starvation is an important, but still understudied, phase of bacterial growth (Finkel, 2006; Kolter, Siegele, & Tormo, 1993; Navarro Lloren, Tormo, & Martinez-Garcia, 2010; Nyström, 2004). It has been reported that 2.3% of *E. coli* genes are associated with this life period according to the GenBank annotation (Tani, Khodursky, Blumenthal, Brown, & Matthews, 2002). CsiD/GlaH is regulated by RpoS, a sigma factor affecting nearly 500 genes or ~10% of the *E. coli* genome (Lacour & Landini, 2004; Patten, Kirchhof, Schertzberg, Morton, & Schellhorn, 2004; Weber, Polen, Heuveling, Wendisch, & Hengge, 2005), and Lrp, a transcription factor modulating more than 400 genes (Tani et al., 2002), with many of the functions for these genes remaining unidentified. As described here, the stationary phase-dependent regulation of *csiD/glaH* and its neighboring genes leads to the production of a complex pathway allowing the organism to metabolize L-lysine as a carbon source.

7. Experimental section

7.1 Strain construction

The Δ*csiD* strain of *E. coli* BL21(DE3) was generated by P1 transduction (Thomason, Constantino, & Court, 2007) using the Δ*csiD* strain of *E.*

coli BW25113 of the Keio collection obtained from the Coli Genetic Stock Center (Baba et al., 2006).

For overproduction of the CsiD protein for biochemical studies, full-length *csiD* was amplified from *E. coli* MG1655 DNA using 5′-CATAAGAGGATCGCTTCATATGAATGCACTGACCG-3′ and 5′-AAGCTTTTACTGATGCGTCTGGTA-3′ primers that introduced *Nde*I and *Hin*dIII sites at the 5′ and 3′ end of the gene, respectively. Ligation of the endonuclease treated PCR product into identically digested pET42b (Novagen) provided an in-frame region encoding CsiD with an *N*-terminal His_6 tag. The resulting plasmid, pET42b-csiD, was transformed into DH5α competent cells, and the amplified plasmid was transformed into *E. coli* BL21(DE3) or the Δ*csiD* cells that were plated on LB agar containing 50 μg/mL of kanamycin.

7.2 Purification of CsiD

A colony of *E. coli* BL21(DE3) pET42b-csiD was used to inoculate a culture of 50 mL LB medium containing 50 μg/mL of kanamycin and this was grown overnight at 37 °C. An aliquot (15 mL) was used to inoculate 1 L of terrific broth (DOT scientific) containing 50 μg/mL kanamycin, and this culture was grown at 37 °C until reaching OD_{600} of 0.2. The culture was shifted to 25 °C and shaken vigorously until reaching OD_{600} of 0.5 at which point the cells were induced with 0.1 mM isopropyl β-D-1-thio-galactopyranoside and the growth continued overnight at 25 °C. The cells were harvested by centrifugation at 7,500 *g* for 10 min at 4 °C. Approximately 3 g of harvested *E. coli* BL21(DE3) pET42b-csiD cells was suspended in 10 mL of lysis buffer containing 20 mM imidazole, 300 mM NaCl, 50 mM Na_2HPO_4, and 1 mM DTT, pH 8. The suspension was sonicated using a Sonifier 450 (Branson) for 1 min at 30 W output power, and 50 % pulse duration. The sonication was repeated four more times with 1 min intervals of cooling in ice. The broken cells were centrifuged at 100,000 *g* for 1 h at 4 °C. The soluble cell-free extracts (~9 mL) were loaded onto an 8 mL Ni-Sepharose 6-Fast Flow column (GE Healthcare), which had been equilibrated with buffer A (20 mM imidazole, 300 mM NaCl, 50 mM Na_2HPO_4, pH 8). The column was washed with buffer B (100 mM imidazole, 300 mM NaCl, 50 mM Na_2HPO_4, pH 8) in order to remove any protein weakly bound to the resin. CsiD was eluted with buffer C (500 mM imidazole, 300 mM NaCl, 50 mM Na_2HPO_4, pH 8). The fractions containing the purified protein were concentrated then passed through a PD-10 desalting column (GE Healthcase) for buffer exchange into 25 mM Tris, 1 mM DTT, pH 8. In some cases (where apoprotein was desired) the CsiD solution was incubated with 1 mM EDTA prior to buffer exchange against the same Tris buffer. Enzyme samples were stored frozen at -80 °C until use. Protein concentrations were determined by using the Bradford assay (Bradford, 1976).

7.3 Product analyses and enzyme assays

In addition to the NMR analysis to monitor glutarate, 2OG, 2HG, and succinate (Fig. 5) and the OPD assay to monitor 2OG (Fig. 6), we measured the consumption of 2OG or glutarate and production of succinate or 2HG by HPLC (HPX-87H organic acids column, Bio-Rad). Reaction mixtures (containing enzyme and reactants in the buffer specified) were incubated for various times at 30 °C, and quenched by adding 5 μL of 6 M sulfuric acid to the 300 μL aliquots. After centrifugation at 10,000 *g* for 5 min the supernatants were centrifuged in spin columns (Amicon ultrafree-MC from Millipore) at 10,000 *g* for 1 min, and 200 μL aliquots were injected onto the organic acids HPLC column that had been previously equilibrated with 13 mM sulfuric acid. The refractive index and absorbance at 214 nm were monitored and the integrated peak intensities were compared to those of authentic standards.

Acknowledgments

We thank Piotr Grzyska, Tina Müller, Dongling Zhao, and Ricardo Aquino for assistance with selected aspects of these studies.

This work was initiated with support by the National Institutes of Health (GM065384 to R.P.H.).

References

Aik, W. S., Chowdhury, R., Clifton, I. J., Hopkinson, R. J., Leissing, T., McDonough, M. A., et al. (2015). Introduction to structural studies on 2-oxoglutarate-dependent oxygenases and related enzymes. In C. J. Schofield, & R. P. Hausinger (Eds.), *2-Oxoglutarate-Dependent oxygenases*. Cambridge, UK: Royal Society of Chemistry.

Baba, T., Ara, T., Hasegawa, M., Takai, Y., Okumura, Y., Baba, M., et al. (2006). Construction of *Escherichia coli* K-12 in-frame, single-gene knockout mutants: The Keio collection. *Molecular and Systems Biology, 2*, 2006.0008.

Bradford, M. M. (1976). A rapid and sensitive method for the quantitation of microgram quantities of protein utilizing the principle of protein-dye binding. *Analytical Biochemistry, 72*, 248–254.

Chance, M. R., Bresnick, A. R., Burley, S. K., Jiang, J.-S., Lima, C. D., Sali, A., et al. (2002). Structural genomics: A pipeline for providing structures for the biochemist. *Protein Science, 11*, 723–738.

Chang, W.-c., Guo, Y., Wang, C., Butch, S. E., Rosenzweig, A. C., Boal, A. K., et al. (2014). Mechanism of the C5 stereoinversion reaction in the biosynthesis of carbapenem antibiotics. *Science, 343*, 1140–1144.

Chang, W.-c., Li, J., Lee, J. L., Cronican, A. A., & Guo, Y. (2016). Mechanistic investigation of a non-heme iron enzyme catalyzed epoxidation in (-)-4'-methoxycyclopenin biosynthesis. *Journal of the American Chemical Society, 138*, 10390–10393.

Cozzani, I., Fazio, A. M., Felici, E., & Barletta, G. (1980). Separation and characterization of NAD- and NADP-specific succinate-semialdehyde dehydrogenase from *Escherichia coli* K-12 3300. *Biochimica et Biophysica Acta, 613*, 309–317.

Dover, S., & Halpern, Y. S. (1974). Genetic analysis of the γ-aminobutyrate utilization pathway in *Escherichia coli* K-12. *Journal of Bacteriology, 117*, 494–501.

Dunham, N. P., Chang, W.-c., Mitchell, A. J., Martinie, R. J., Zhang, B., Bergman, J. A., et al. (2018). Two distinct mechanisms for C-C desaturation by iron(II)- and 2-(oxo)glutarate-dependent oxygenases: Importance of α-heteroatom assistance. *Journal of the American Chemical Society, 140*, 7116–7126.

Eichhorn, E., van der Ploeg, J. R., Kertesz, M. A., & Leisinger, T. (1997). Characterization of α-ketoglutarate-dependent taurine dioxygenase from *Escherichia coli*. *Journal of Biological Chemistry, 272*, 23031–23036.

Fang, D., & Cisneros, G. A. (2013). Alternative pathway for the reaction catalyzed by DNA dealkylase AlkB from ab initio QM/MM calculations. *Journal of Chemical Theory and Computation, 10*, 5136–5148.

Finkel, S. E. (2006). Long-term survival during stationary phase: Evolution and the GASP phenotype. *Nature Reviews in Microbiology, 4*, 113–120.

Fukumori, F., & Hausinger, R. P. (1993). Purification and characterization of 2,4-dichlorophenoxyacetate/α-ketoglutarate dioxygenase. *Journal of Biological Chemistry, 268*, 24311–24317.

Galonić, D. P., Barr, E. W., Walsh, C. T., Bollinger, J. M., Jr., & Krebs, C. (2007). Two interconverting Fe(IV) intermediates in aliphatic chlorination by the halogenase CytC3. *Nature Chemical Biology, 3*, 113–116.

Grzyska, P. K., Appelman, E. H., Hausinger, R. P., & Proshlyakov, D. A. (2010). Insight into the mechanism of an iron dioxygenase by resolution of steps following the Fe(IV)=O species. *Proceedings of the National Academy of Sciences, 107*, 3982–3987.

Hausinger, R. P. (2004). Fe(II)/α-ketoglutarate-dependent hydroxylases and related enzymes. *Critical Reviews in Biochemistry and Molecular Biology, 39*, 21–68.

Hausinger, R. P. (2015). Biochemical diversity of 2-oxoglutarate-dependent oxygenases. In C. J. Schofield, & R. P. Hausinger (Eds.), *2-Oxoglutarate-Dependent oxygenases* (pp. 1–58). Cambridge, UK: Royal Society of Chemistry.

Hayashi, T., Tsuchiya, H., Todoriki, H., & Naruse, H. (1982). High-performance liquid chromatographic determination of alpha-keto acids in human urine and plasma. *Analytical Biochemistry, 122*, 173–179.

Hegg, E. L., Whiting, A. K., Saari, R. E., McCracken, J., Hausinger, R. P., & Que, L., Jr. (1999). Herbicide-degrading α-keto acid-dependent enzyme TfdA: Metal coordination environment and mechanistic insights. *Biochemistry, 38*, 16714–16726.

Hengge, R. (2008). The two-component network and the general stress sigma factor Rpo(σ^s) in *Escherichia coli*. *Advances in Experimental Medicine & Biology, 631*, 40–53.

Herr, C. Q., & Hausinger, R. P. (2018). Amazing diversity in biochemical roles of Fe(II)/2-oxoglutarate oxygenases. *Trends in Biochemical Sciences, 43*, 517–532.

Hoffart, L. M., Barr, E. W., Guyer, R. B., Bollinger, J. M., Jr., & Krebs, C. (2006). Direct spectroscopic detection of a C-H-cleaving high spin Fe(IV) complex in a prolyl-4-hydroxylase. *Proceedings of the National Academy of Sciences, 103*, 14738–14743.

Islam, M. S., Leissing, T. M., Chowdhury, R., Hopkinson, R. J., & Schofield, C. J. (2018). 2-Oxoglutarate-dependent oxygenases. *Annual Review of Biochemistry, 87*, 585–620.

Jorge, J. M. P., Perez-Garcia, F., & Wendisch, V. F. (2017). Anew metabolic route of the fermentative production of 5-aminovalerate from glucose and alternative carbon sources. *Bioresource Technology, 245*, 1701–1709.

Kahnert, A., & Kertesz, M. A. (2000). Characterization of a sulfur-regulated oxygenative alkylsulfatase from *Pseudomonas putida* S-313. *Journal of Biological Chemistry, 275*, 31661–31667.

Kalliri, E., & Hausinger, R. P. (2008). Identification of *Escherichia coli* YgaF as an L-2-hydroxyglutarate oxidase. *Journal of Bacteriology, 190*, 3793–3798.

Kanehisa, M., Goto, S., Sato, Y., Kawashima, M., Furumichi, M., & Tanabe, M. (2014). Data, information, knowledge and principle: Back to metabolism in KEGG. *Nucleic Acids Research, 42*, D199–D205.

Knorr, S., Sinn, M., Galetskiy, D., Williams, R. M., Wang, C., Müller, N., et al. (2018). Widespread bacterial lysine degradation proceeding via glutarate and *L*-2-hydroxyglutarate. *Nature Communications, 9*, 5071.

Kolter, R., Siegele, D. A., & Tormo, A. (1993). The stationary phase of the bacterial life cycle. *Annual Review of Microbiology, 47*, 855–874.

Lacour, S., & Landini, P. (2004). σ^s-Dependent gene expression at the onset of stationary phase in *Escherichia coli*: Function of σ^s-dependent genes and identification of their promoter sequences. *Journal of Bacteriology, 186*, 7186–7195.

Lemonnier, M., & Lane, D. (1998). Expression of the second lysine decarboxylase gene of *Escherichia coli*. *Microbiology, 144*, 751–760.

Liu, P., Zhang, H., Lv, M., Hu, M., Li, Z., Gao, C., et al. (2014). Enzymatic production of 5-aminovalerate from L-lysine using L-lysine monooxygenase and 5-aminovaleramide amidohydrolase. *Scientific Reports, 4*(5657).

Lohkamp, B., & Dobritzsch, D. (2008). A mixture of fortunes: The curious determination of the structure of *Escherichia coli* BL21 Gab protein. *Acta Crystallographica, D64*, 407–415.

Marschall, C., Labrousse, V., Kreimer, M., Wichart, D., Kolb, A., & Hengge-Aronis, R. (1998). Molecular analysis of the regulation of *csiD*, a carbon starvation-inducible gene in *Escherichia coli* that is exclusively dependent on σ^s and requires activation by cAMP-CRP. *Journal of Molecular Biology, 276*, 339–353.

Martinez, S., & Hausinger, R. P. (2015). Catalytic mechanisms of Fe(II)- and 2-oxoglutarate-dependent oxygenases. *Journal of Biological Chemistry, 290*, 20702–20711.

Martinez, S., & Hausinger, R. P. (2016). Biochemical and spectroscopic characterization of the non-heme Fe(II)- and 2-oxoglutarate-dependent ethylene-forming enzyme from *Pseudomonas syringae* pv. *phaseolicola* PK2. *Biochemistry, 55*, 5989–5999.

Matthews, M. M., Krest, C., Barr, E. W., Vaillancourt, F., Walsh, C. T., Green, M., et al. (2009). Substrate-triggered formation and remarkable stability of the C-H-cleaving chloroferryl intermediate in the aliphatic halogenase, SyrB2. *Biochemistry, 48*, 4331–4343.

Metzner, M., Germer, J., & Hengge, R. (2004). Multiple stress signal integration in the regulation of the complex σ^s-dependent *csiD-ygaF-gabDTP* operon in *Escherichia coli*. *Molecular Microbiology, 51*, 799–811.

Mitchell, A. J., Dunham, N. P., Martinie, R. J., Bergman, J. A., Pollock, C. J., Hu, K., et al. (2017). Visualizing the reaction cycle in an iron(II)- and 2-(oxo)-glutarate-dependent hydroxylase. *Journal of the American Chemical Society, 139*, 13830–13836.

Navarro Lloren, J. M., Tormo, A., & Martinez-Garcia, E. (2010). Stationary phase in gram-negative bacteria. *FEMS Microbiology Reviews, 34*, 476–495.

Niegemann, E., Schulz, A., & Bartsch, K. (1993). Molecular organization of the *Escherichia coli gab* cluster: Nucleotide sequence of the structural genes *gabD* and *gabP* and expression of the GABA permease gene. *Archives of Microbiology, 160*, 454–460.

Nyström, T. (2004). Stationary-phase physiology. *Annual Review of Microbiology, 58*, 161–181.

Patten, C. L., Kirchhof, M. G., Schertzberg, M. R., Morton, R. A., & Schellhorn, H. E. (2004). Microarray analysis of RpoS-mediated expression in *Escherichia coli* K-12. *Molecular and General Genetics, 272*, 580–591.

Pavel, E. G., Zhou, J., Busby, R. W., Gunsior, M., Townsend, C. A., & Solomon, E. I. (1998). Circular dichroism and magnetic circular dichroism spectroscopic studies of the non-heme ferrous active site in clavaminate synthase and its interaction with α-ketoglutarate cosubstrate. *Journal of the American Chemical Society, 120*, 743–753.

Price, J. C., Barr, E. W., Glass, T. E., Krebs, C., & Bollinger, J. M., Jr. (2003). Evidence for hydrogen abstraction from C1 of taurine by the high-spin Fe(IV) intermediate detected during oxygen activation by taurine:α-ketoglutarate dioxygenase (TauD). *Journal of the American Chemical Society, 125*, 13008–13009.

Price, J. C., Barr, E. W., Tirupati, B., Bollinger, J. M., Jr., & Krebs, C. (2003). The first direct characterization of a high-valent iron intermediate in the reaction of an α-ketoglutarate-dependent dioxygenase: A high-spin Fe(IV) complex in taurine/α-ketoglutarate dioxygenase (TauD) from *Escherichia coli*. *Biochemistry, 42*, 7497–7508.

Proshlyakov, D. A., Henshaw, T. F., Monterosso, G. R., Ryle, M. J., & Hausinger, R. P. (2004). Direct detection of oxygen intermediates in the non-heme Fe enzyme taurine/α-ketoglutarate dioxygenase. *Journal of the American Chemical Society, 126*, 1022–1023.

Proshlyakov, D. A., McCracken, J., & Hausinger, R. P. (2017). Spectroscopic analyses of 2-oxoglutarate-dependent oxygenases: TauD as a case study. *Journal of Biological Inorganic Chemistry, 22*, 367–379.

Revelles, O., Espinosa-Urgel, M., Molin, S., & Ramos, J. L. (2004). The *davDT* operon of *Pseudomonas putida*, involved in lysine catabolism, is induced in response to the pathway intermediate δ-aminovaleric acid. *Journal of Bacteriology, 186*, 3439–3446.

Ryle, M. J., Padmakumar, R., & Hausinger, R. P. (1999). Stopped-flow kinetic analysis of *Escherichia coli* taurine/α-ketoglutarate dioxygenase: Interactions with α-ketoglutarate, taurine, and oxygen. *Biochemistry, 38*, 15278–15286.

Sabo, D. L., Boeker, E. A., Byers, B., Waron, H., & Fischer, E. H. (1974). Purification and physical properties of inducible *Escherichia coli* lysine decarboxylase. *Biochemistry, 13*, 662–670.

Samsonova, N. N., Smirnov, S. V., Altman, I. B., & Ptitsyn, L. R. (2003). Molecular cloning and characterization of *Escherichia coli* K12 *ygjG* gene. *BMC Microbiology, 3*, 2.

Schneider, B. L., Ruback, S., Kiupakis, A. K., Kasbarian, H., Pybus, C., & Reitzer, L. (2002). The *Escherichia coli gabDTPC* operon: Specific γ-aminobutyrate catabolism and nonspecific induction. *Journal of Bacteriology, 184*, 6976–6986.

Schofield, C. J., & Hausinger, R. P. (Eds.). (2015). *2-Oxoglutarate-dependent oxygenases*. Cambridge, UK: Royal Society of Chemistry.

Tani, T. H., Khodursky, A., Blumenthal, R. M., Brown, P. O., & Matthews, R. G. (2002). Adaptation to famine: A family of stationary-phase genes revealed by microarray analysis. *Proceedings of the National Academy of Sciences, 99*, 13471–13476.

Thomason, L. C., Constantino, N., & Court, D. L. (2007). *E. coli* genome manipulation by P1 transduction. *Current Protocols in Molecular Biology, 79*, 1.17.11-11.17.18.

Thompson, M. G., Blake-Hedges, J. M., Cruz-Morales, P., Barajas, J. F., Curran, S. C., Eiben, C. B., Harris, N. C., Benites, V. T., Gin, J. W., Sharpless, W. A., Twigg, F. F., Skyrud, W., Krishna, R. N., Pereira, J. H., Baidoo, E. E. K., Petzold, C. J., Adams, P. D., Arkin, A. P., Deutschbauer, A. M., & Keasling, J. D. (2019). Massively parallel fitness profiling reveals multiple novel enzymes in Pseudomonas putida lysine metabolism. *mBio, 10*, e02577–18.

Weber, H., Polen, T., Heuveling, J., Wendisch, V. F., & Hengge, R. (2005). Genome-wide analysis of the general stress response network in *Escherichia coli*: σ^S-dependent genes, promoters, and sigma factor selectivity. *Journal of Bacteriology, 187*, 1591–1603.

Weichart, D., Lange, R., Henneberg, N., & Hengge-Aronis, R. (1993). Identification and characterization of stationary phase-inducible genes in *Escherichia coli*. *Molecular Microbiology, 10*, 407–420.
van der Werf, M. J., Overkamp, K. M., Muilwijk, B., Coulier, L., & Hankemeier, T. (2007). Microbial metabolomics: Toward a platform with full metabolome coverage. *Analytical Biochemistry, 370*, 17–25.
Wu, L.-F., Meng, S., & Tang, G.-L. (2016). Ferrous iron and α-ketoglutarate dependent dioxygenases in the biosynthesis of microbial natural products. *Biochimica et Biophysica Acta, 1864*, 453–470.
Zhang, M., Gao, C., Guo, X., Guo, S., Kang, Z., Xiao, D., et al. (2018). Increased glutarate production by blocking the glutaryl-CoA dehydrogenation pathway and a catabolic pathway involving L-2-hydroxylglutarate. *Nature Communications, 9*, 2114.

CHAPTER THREE

Comparison of DNA and RNA substrate effects on TET2 structure

Emmett M. Leddin and G. Andrés Cisneros*
Department of Chemistry, University of North Texas, Denton, TX, United States
*Corresponding author: E-mail: Gerardo.Cisneros@unt.edu

Contents

1. Introduction 92
2. Methodology 94
   2.1 Structure preparation and parametrization 94
   2.2 MD simulations 95
   2.3 Comparison data 95
   2.4 Analysis 96
3. Results and discussion 96
   3.1 Backbone analysis 98
   3.2 Correlation matrices 101
   3.3 Hydrogen bond analysis 103
   3.4 Energy decomposition analysis 105
4. Conclusions 109
Acknowledgments 109
References 109

Abstract

Ten-eleven translocation (TET) enzymes can perform the stepwise oxidation of 5-methylcytosine (5mC) to 5-carboxylcytosine on both single-stranded (ss) and double-stranded (ds) DNA and RNA. It has been established that TET2 has a preference for ds DNA substrates, but it can catalyze the oxidation reaction on both ssDNA and RNA. The reasons for this substrate preference have been investigated for only a substrate 5mC ribonucleotide in a DNA strand, but not other nucleic acid configurations (*Biochemistry* **58** (2019) 411). We performed molecular dynamics simulations on TET2 with various ss and ds substrates in order to better understand the structural and dynamical reasons for TET2's preference to act on ds DNA. Our simulations show that substrates that have a ribonucleotide experience several disruptions in their overall

Advances in Protein Chemistry and Structural Biology, Volume 117
ISSN 1876-1623
https://doi.org/10.1016/bs.apcsb.2019.05.002
© 2019 Elsevier Inc.
All rights reserved.

backbone shape, hydrogen bonding character, and non−bonded interactions. These differences appear to lead to the instability of ribonucleotide in the active site, and provide further rational for TET2's experimental behavior.

1. Introduction

The methylation and demethylation of DNA and RNA is important for many biological functions. Nucleobase methylation is a mainstay epigenetic feature, and it plays a role in tasks such as X-chromosome inactivation and genomic imprinting (Jones, 2012; Kohli & Zhang, 2013; Rasmussen & Helin, 2016; Williams, Christensen, & Helin, 2012).

Most epigenetic methylation marks on DNA occur as 5-methylcytosine (5mC), whereas most RNA epigenetic methylation marks occur as N6-methyladenosine (Chen-Kiang, Nevins, & Darnell, 1979; Jin, Li, & Robertson, 2011). DNA methylation is catalyzed by DNA methyltransferases (DNMTs) acting on deoxycytidine and adenosine, respectively (Bird, 2002).

Several enzymes in the α-ketoglutarate- (α-KG) and Fe(II)-dependent superfamily can perform nucleoside demethylation. The general catalytic cycle for the oxidation of alkyl moieties can be separated into two general stages. The first stage involves the binding of the substrate and co−factors, α-KG and O_2, and formation of a ferryl (Fe(IV)=O) intermediate with the concomitant oxidation of α-KG to succinate and CO_2. Subsequently, the second stage involves the oxidation of the methyl moiety on the substrate. Computational simulations using hybrid quantum mechanics/molecular mechanics (QM/MM) have identified several features of the catalytic cycle for TET2 and a related enzyme, AlkB (Fang & Cisneros, 2014; Fang, Lord, & Cisneros, 2013; Torabifard & Cisneros, 2018).

Ten eleven translocation (TET) dioxygenases belong to the α-KG- and Fe(II)-dependent superfamily of enzymes. A key structural feature of these enzymes is a conserved double sheet β-helix, also known a jelly-roll fold, in their catalytic cores (Aik, McDonough, Thalhammer, Chowdhury, & Schofield, 2012). The TET name derives from a common translocation between the *TET1* and *MLL* genes (Lorsbach et al., 2003). TET enzymes catalyze the stepwise oxidation of 5mC to 5-hydroxymethylcytosine (5hmC), 5-carboxylcytosine (5caC), and 5-formylcytosine (5fC), sequentially (Fig. 1). These enzymes can perform this oxidation on both methylated DNA and RNA nucleosides, as part of single-stranded (ss) or double-

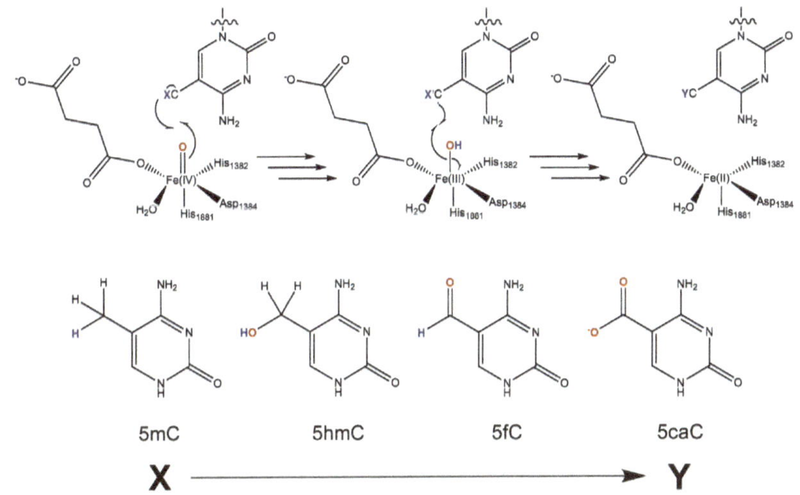

Fig. 1 The proposed reaction mechanism for TET enzymes. The conversion from 5mC to 5caC occurs by stepwise oxidation, where the initial reactant "X," is converted into its immediate product, "Y."

stranded (ds) substrates (Fu et al., 2014; Kizaki & Sugiyama, 2014). A part of the reaction catalyzed by TET enzymes involves a base flipping (eversion) of the nucleic acid, wherein the target residue is flipped out from the substrate strand into the active site (Hong & Cheng, 2016). Base flipping is not limited to TET enzymes, but rather, it is seen across several enzymes that read, write, or otherwise modify DNA bases, including the DNA glycosylases involved in base excision repair (BER) (Brooks, Adhikary, Rubinson, & Eichman, 2013; Drohat & Coey, 2016). The enzymes involved with BER complete the final conversion of the 5caC oxidation product back into dC.

Two of the three TET enzymes, TET1 and TET3, contain a CXXC domain that can bind directly to DNA (Rasmussen & Helin, 2016). While not required for oxidation of 5mC, the CXXC domain has been proposed to increase TET's selectivity (Shen, Song, He, & Zhang, 2014). TET2 does not contain a CXXC binding domain, but is known to partner with IDAX, which does (Ko et al., 2013; Kohli & Zhang, 2013). Recently, a scaffold in the active site of TET2, composed by T1372 and Y1902, has been shown to be required for the sequential oxidation of the substrate (Liu et al., 2016).

It has been shown that TET2 recognizes the cytosine guanine (CpG) dinucleotide sequence of DNA and is unlikely to discriminate between 5mC, 5hmC, and 5fC in the active site (Hu et al., 2015; Tsai & Tainer, 2013). TET2's inability to discern between reactants is proposed to result from the cytosine base adopting nearly identical orientations in the active site. Despite their similarities in orientation, 5mC and 5hmC have been shown to stabilize DNA duplexes (Thalhammer, Hansen, El-Sagheer, Brown, & Schofield, 2011), whereas DNA duplexes containing either 5fC or 5caC destabilize DNA duplexes (Raiber et al., 2015), although this has been challenged (Hardwick et al., 2017). This destabilization leads to what has been deemed F-form DNA, and reduction from 5fC to 5hmC can return the backbone structure to its typical B-form.

While TET2 has been shown to act on both DNA and RNA substrates, the differences between both contexts is poorly understood. Recently, we identified a distinct pattern for selectivity of TET2 complexes, favoring dsDNA and ssDNA, and disfavoring dsRNA and ssRNA (DeNizio, Liu, Leddin, Cisneros, & Kohli, 2019). Through comparing with previous computational work on TET2-DNA complexes (Liu et al., 2016), we uncovered several residues with large differences in their non—bonded interactions with the target residue, as well as a shift in the DNA backbone structure with the inclusion of an RNA base.

In this work, we extend our computational work to further investigate how a variety of different RNA and DNA substrates impact TET2 complexes and provide a possible explanation for the pattern of selectivity seen experimentally.

2. Methodology

2.1 Structure preparation and parametrization

The crystal structure of a TET2-DNA complex (PDBID: 4NM6) (Hu et al., 2013) was prepared as previously described (DeNizio et al., 2019; Liu et al., 2016). Briefly, the simulation systems involve the following features: a magnesium surrogate for the active site iron, protonation of the His residues coordinating the active site divalent cation on the δ nitrogen, and the insertion of a glycine-serine linker. The linker was introduced using the Modeller program (Šali & Blundell, 1993; Salomon-Ferrer, Götz, Poole, Le Grand, & Walker, 2013).

The 5hmrC, 5carC, and 5frC ribonucleotide parameters were all generated using PyRED (Bayly, Cieplak, Cornell, & Kollman, 1993; Dupradeau et al., 2010; Vanquelef et al., 2011; Wang, Becker, Cieplack, & Dupradeau, 2013). The published parameters for deoxyribonucleotide cytosine derivatives, α-KG, Mg(II), and Zn(II) were used (Liu et al.,

2016), in addition to the 5mrC parameters from our previous work (DeNizio et al., 2019).

2.2 MD simulations

The pmemd.cuda program from the AMBER16 suite (Case et al., 2017; Götz et al., 2012; Salomon-Ferrer et al., 2013) was used to perform molecular dynamics (MD) simulations for all TET2-nucleic acid complexes. Eight systems were created and simulated with the ff99SB force field. The systems considered include: 1) ssDNA with 5mrC (ss 5mrC); 2) ssDNA with 5mdC (ss 5mdC); 3) dsDNA with 5hmrC, 4) 5carC, or 5) 5frC in the substrate strand with 5mdC in the complementary strand (respectively 5hmrC:5mdC, 5carC:5mdC, and 5frC:5mdC); 6) dsDNA with 5mrC in both the substrate and complementary strands (5mrC:5mrC); 7) dsDNA with 5mdC in the substrate strand and 5mrC in the complementary strand (5mdC:5mrC); and 8) dsDNA with 5mdC in the substrate and dC in the complementary strand (5mdC:dC).

The LEaP module in AMBER was used to generate each system (Schafmeister, Ross, & Romanovski, 1995). Each system was neutralized to a net charge of zero using potassium ions and solvated with TIP3P explicit water extending at least 12Å from the system surface in a truncated octahedron (Jorgensen, Chandrasekhar, Madura, Impey, & Klein, 1983). The smooth particle mesh Ewald method with a 9Å direct space cutoff was used to treat long-range electrostatic interactions (Essmann et al., 1995). The same cutoff was used for the Van der Waals interactions.

A 2 fs time step was used for simulations with the Berendesen thermostat in the *NVT* ensemble (Berendsen, Postma, van Gunsteren, DiNola, & Haak, 1984). A minimum of three replicates for each system were run for 90 ns of production, resulting in 270 ns of total production time per system, for a total of over 2 s of production simulation time.

2.3 Comparison data

Earlier work studied dsDNA with both 5mdC in substrate and complement (5mdC:5mdC) (Liu et al., 2016). These simulations were minimized and heated in the *NVT* ensemble and run for 40 ns of production using the *NPT* ensemble. Later work looked at dsDNA with 5mrC in the substrate and complement (5mrC:5mdC) using the *NVT* ensemble procedure described here, as well as comparing both procedures to ensure meaningful comparisons could be made (DeNizio et al., 2019).

2.4 Analysis

Different structural analyses, including hydrogen bond analysis, root mean square deviations (RMSDs), root mean square fluctuations (RMSFs), matrix correlations, and nucleic acid backbone torsions and base pair geometries were performed using cpptraj (Babcock, Pednault, & Olson, 1994; Olson et al., 2001; Roe & Cheatham, 2013). Averaging for these analyses was performed using the data.table, abind, and tidyverse packages in R (Dowle & Srinivasan, 2018; Plate & Heiberger, 2016; R Core Team, 2018; Wickham, 2017). VMD and UCSF Chimera were used for data visualization and image creation (Humphrey, Dalke, & Schulten, 1996; Pettersen et al., 2004) An in-house FORTRAN90 program was used to perform an energy decomposition analysis (EDA) (Dewage & Cisneros, 2015; Elias & Cisneros, 2014; Graham, Syeda, & Cisneros, 2012). The non−bonded interactions, either Coulomb or Van der Waals, for each individual residue, i, are computed with every other residue j as per (Eq. 1).

$$E_{int} = < E_i - E_j >, \tag{1}$$

where the individual non−bonded components, Coulomb and Van der Waals, for every residue pair i and j are averaged over the generated ensemble, and $< \cdots >$ denotes an ensemble average.

The sum of the two non−bonded components for each individual E_{int} values provides a qualitative understanding of the non−bonded interactions for every residue pair (Eq. 2).

$$E_{Tot}^{Non-bond} = E_{int,Coul} + E_{int,VdW} \tag{2}$$

Systems can be compared relative to each other by taking the difference in the total non−bonded interactions between two systems (Eq. 3).

$$\Delta E_{Tot}^{Non-bond} = E_{Tot,ref}^{Non-bond} - E_{Tot}^{Non-bond} \tag{3}$$

3. Results and discussion

Investigating various nucleosides in a TET2-DNA complex could lead to a better understanding of TET2's preference for (a) ds over ss substrates and (b) DNA over RNA. Because both DNA and RNA undergo methylation to 5mC, it is important to understand if or how the interactions between TET2 and different DNA or RNA substrates change.

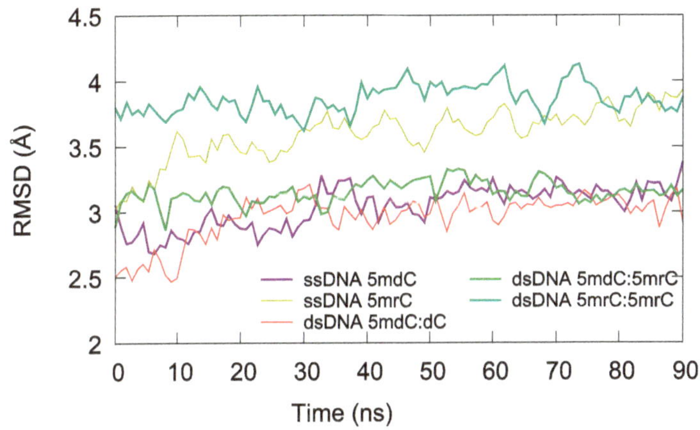

Fig. 2 RMSD plots for the ss 5mdC, ss 5mrC, 5mdC:dC, 5mdC:5mrC, and 5mrC:5mrC systems over 90 ns of production. Bézier line smoothing has been applied.

The RMSD analysis for each system is shown in Figs. 2 and 3. The ss 5mrC and 5mrC:5mrC systems showed slightly higher overall deviation from the crystal structure than the other systems studied. This may due to the fact that the lack of the complementary strand reduced the number of contacts between the substrate and the protein resulting in higher flexibility of the ssDNA strand in these systems (vide infra).

The RMSF analysis suggests that the overall fluctuation of most protein residues fluctuates similarly between the deoxy and ribo systems. Conversely, the nucleic acids exhibit larger fluctuations in the ribo systems

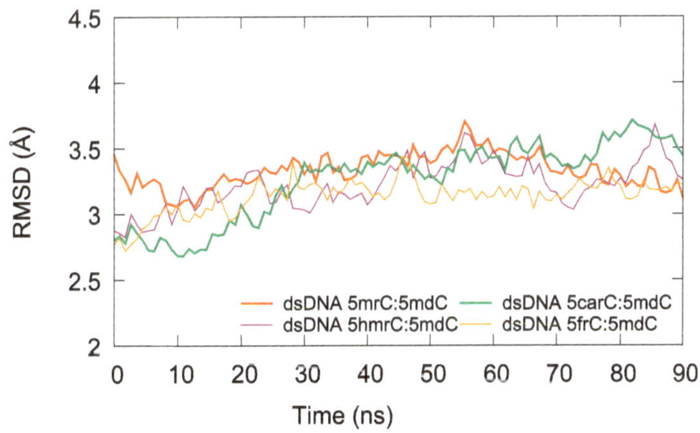

Fig. 3 RMSD plots for the 5mrC:5mdC, 5hmrC:5mdC, 5carc:5mdC, and 5frC:5mdC systems over 90 ns of production. Bézier line smoothing has been applied.

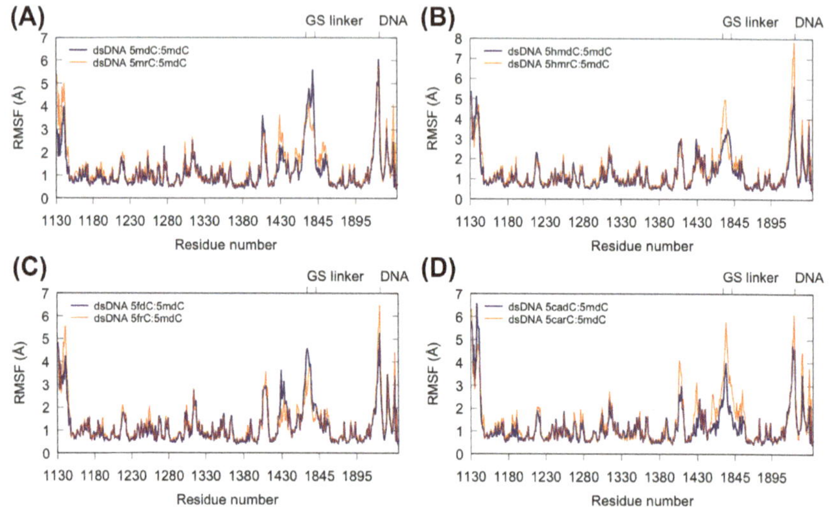

Fig. 4 RMSF plots for the (A) 5mrC:5mdC, (B) 5hmrC:5mdC, (C) 5carc:5mdC, and (D) 5frC:5mdC systems. The region with the glycine-serine linker is denoted by "GS linker."

(Fig. 4). This difference suggests that the inclusion of the ribonucleotide creates an instability in the nucleic acid backbone.

3.1 Backbone analysis

In our previous work with the 5mrC:5mdC system, we identified a slight shift in the DNA substrate from B-form to A-form, due to the inclusion of the ribonucleotide (DeNizio et al., 2019). To see if a similar pattern emerged for the other systems considered herein, we looked again at the different χ and δ angles throughout the course of the simulations. Many systems showed distinct shifts in the substrate's backbone angle populations in the ribo systems, when compared to their deoxy counterparts (Fig. 5). Remarkably, the 5fdC:5mdC and 5frC:5mdC systems have the same spread of values for χ and δ as 5mrC:5mdC. This suggests that the positioning of the target nucleic acid in the active site is more similar between 5fdC:5mdC and 5mrC:5mdC than 5mdC:5mdC.

The structure of the DNA helix is impacted in the backbone arrangement and base pairing by the identity of both complement and substrate nucleobases (Packer & Hunter, 1998). Here, we found that changing the composition of, or removing, the complementary strand impacted the substrate strand's backbone angles (Fig. 6). The ssDNA system shared its distinct populations for 5mdC and 5mrC with the dsDNA 5mdC:5mdC and 5mrC:

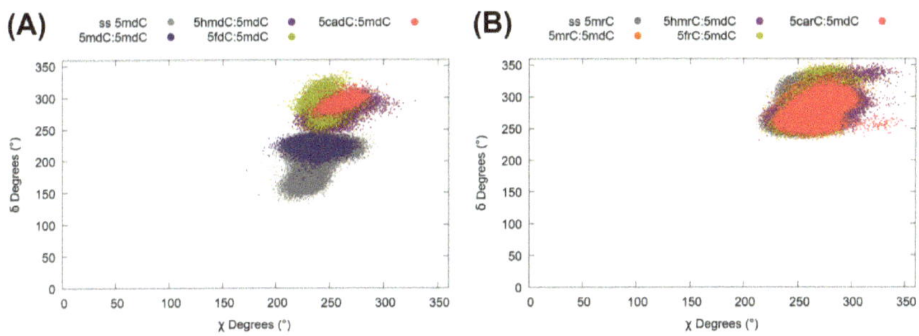

Fig. 5 δ and χ angle populations for the target substrate base in the deoxy systems (A) compared with the ribo systems (B). The 5mdC:5mdC system in (A) is blue and the 5mrC:5mdC system in (B) is orange. In both subfigures, the different 5hmC systems are purple, 5fC systems are gold, and 5caC systems are pink. All replicates are shown, excluding one from 5carC:5mdC exhibiting artifacting.

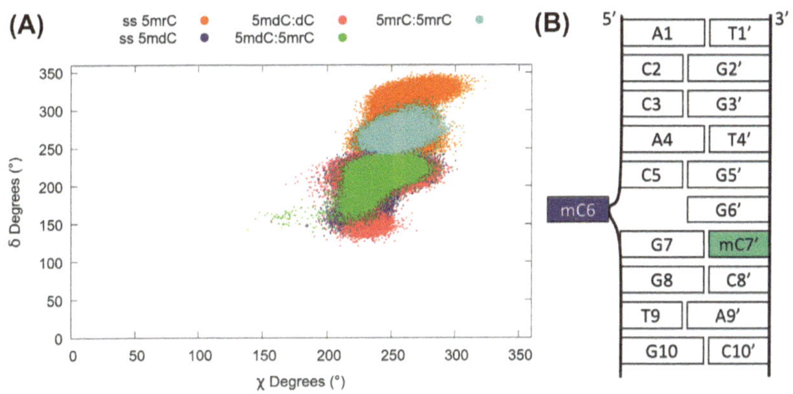

Fig. 6 (A) δ and χ angle populations for the target substrate base in the systems with altered complementary strands: 5mdC:dC (pink), 5mdC:5mrC (green), and 5mdC:5mdC (cyan) systems. (B) The original DNA strand, showing the altered residues for substrate (mC6, blue) and complement (mC7′, sea green). All replicates are shown.

5mdC systems, respectively. While changing the complementary strand can change the substrate's backbone angles, the complete removal of the complementary strand impacts the backbone stability, and thus the spread of the χ and δ angle populations.

The differences in the backbone and base–pairing arrangements can also by observed by the changes to the shift and slide parameters for the C3:G3′ and A4:T4' base pairs (Fig. 7). Each of the ribo systems with a 5mdC complement strand have a wider spread for these values. Alterations to the complementary strand tightened the range for shift and slide, where 5mdC:dC most closely resembled 5mdC:5mdC (Fig. 7E).

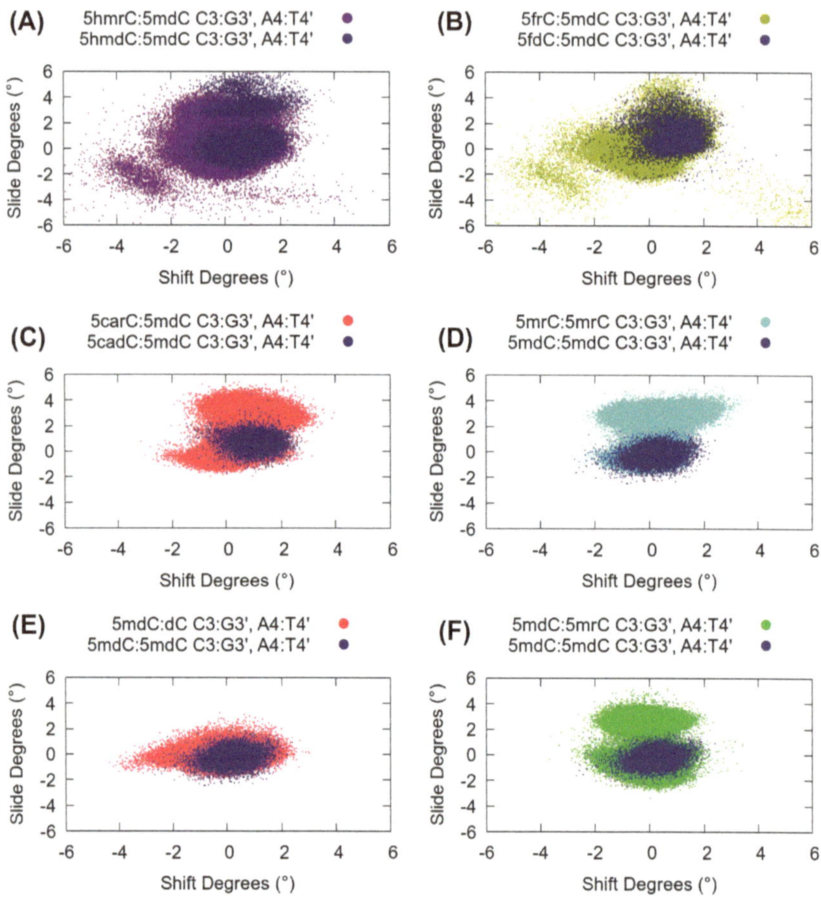

Fig. 7 Differences in shift and slide for the C3:G3′ and A4:T4′ base pairs for (A) 5hmrC:5mdC (purple), (B) 5frC:5mdC (gold), (C) 5carC:5mdC (pink), (D) 5mrC:5mrC (cyan), (E) 5mdC:dC (pink), and (F) 5mdC:5mrC (green). In (A)–(C), the blue corresponds to the deoxy reference system; in (D)–(F), the blue corresponds to 5mdC:5mdC. All replicates, excluding one from 5carC:5mdC with incalculable values, are shown.

As discussed in the introduction, Raiber et al. (2015) uncovered a distinct difference between 5fdC and the other oxidation products, wherein the backbone adopts a structure that they have termed F-DNA. These findings have been challenged by Hardwick et al. (2017). Our simulations show that the backbone parameters for the 5mrC:5mdC system most closely resemble those from the 5fdC:5mdC and 5frC:5mdC systems. Since this system shifted away from B-form DNA, but does not fit entirely with the definition of A-form DNA, the modified nucleic acid strands exhibit characteristics from A-, B-, and F-form DNA, which may help explain this discrepancy.

3.2 Correlation matrices

Correlation matrices show areas of correlation, anti-correlation, or no correlation for a protein relative to itself. Each of the ds systems simulated here show strong similarities in their self-correlations (Fig. 8).

Building from this, taking the difference of the self-correlations between two systems can provide insight into their dynamical differences. Differences in the correlations between ss 5mdC and ss 5mrC show that the ss 5mrC system has slightly stronger correlation between the DNA-interacting loops (residues 1256-1273 and 1288-1312), but slightly stronger anti-correlation with the rest of the protein (Fig. 9). The ss 5mdC/ss 5mrC difference was nearly identical in magnitude for the comparison between 5fdC:5mdC and 5frC:5mdC. The inherent flexibility of the ssDNA, and the aforementioned conformational difference of ds helices with 5fdC, together may explain the similarities in the correlated and anticorrelated movements of the four systems.

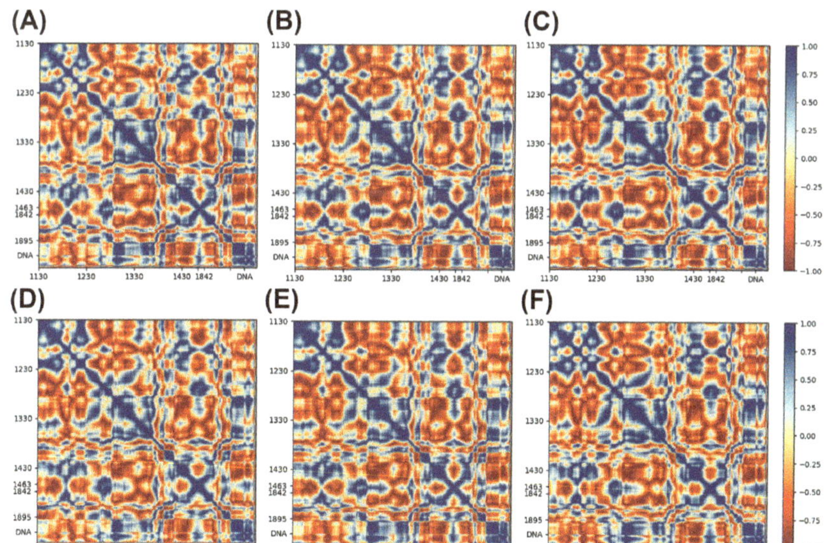

Fig. 8 Self-correlation matrices for (A) 5hmrC:5mdC, (B) 5frC:5mdC, (C) 5carC:5mdC, (D) 5mdC:dC, (E) 5mdC:5mrC, and (F) 5mrC:5mrC. Correlated movement is shown in blue, anti-correlated movement is shown in red, and areas with no correlation are yellow.

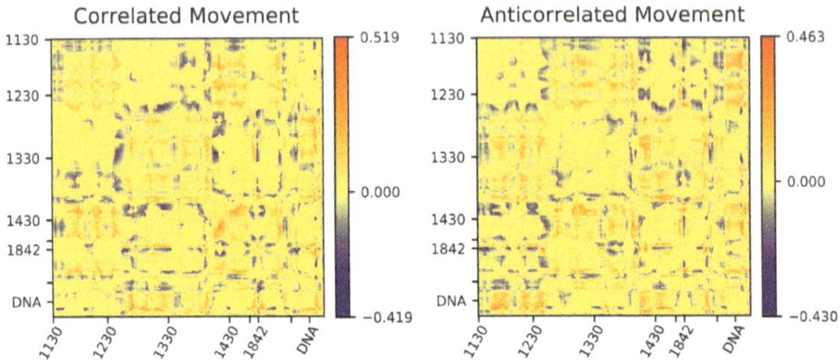

Fig. 9 Differences in the correlation matrices between ssDNA systems. Areas colored blue are stronger in ss 5mdC, whereas areas colored orange are stronger in the ss 5mrC system.

Smaller differences are observed between the 5hmdC:5mdC and 5mrC:5mdC systems (Fig. 10). Compared with the ssDNA systems, the DNA-interacting loops (residues 1256-1273 and 1288-1312) showed slightly stronger correlation between the DNA-interacting loops in the 5hmdC:5mdC system. Comparing each system with an altered complementary strand to the 5mdC:5mdC system showed the most dramatic differences in 5mdC:dC and 5mrC:5mrC, with 5mdC:5mrC exhibiting the same differences more subtly. Each of these systems was less strongly correlated in the DNA-interacting loops than the reference. Taken together, these results suggest that the movements of the DNA-interacting region differ with the different modified cytosine bases, and that the extent of these differences is dependent on both the substrate and complement (or lack thereof) strands.

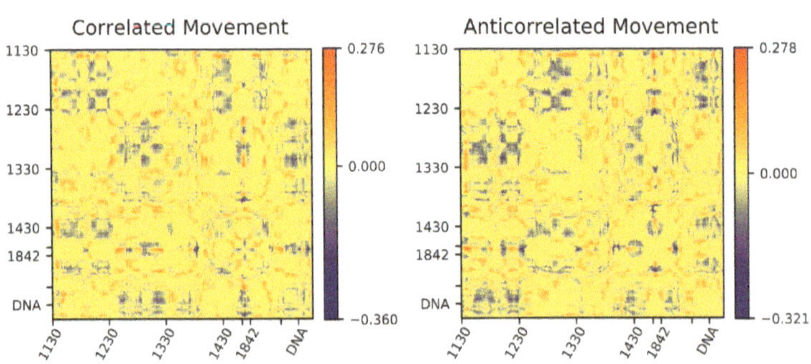

Fig. 10 Differences in the correlation matrices between 5hmdC:5mdC and 5mrC:5mdC systems. Areas colored blue are stronger in 5hmdC:5mdC system, whereas areas colored orange are stronger in the 5mrC:5mdC system.

3.3 Hydrogen bond analysis

The replacement of a deoxy substrate by a ribo substrate induced several changes to the protein's hydrogen bonding interactions. When compared to their respective deoxy systems, both 5hmrC:5mdC and 5carC:5mdC had hydrogen bond between α-KG and R1261 for less time overall, whereas 5frC:5mdC had an increased length of this interaction. A similar pattern, where the hydrogen bond increased in the 5frC:5mdC system and decreased in the 5hmrC:5mdC and 5carC:5mdC systems, was also seen between D1384 and N1387. D1384 is coordinated to the divalent metal center, while N1387 and R1261 both surround the catalytic core (Fig. 11A). Because of this, changes to their orientations induced by the ribo oxygen could impact the demethylation reaction, and thus impede activity.

Both 5hmdC and 5fdC are known to form hydrogen bonds between H1904 and N1387 as part of the stabilization for the flipped in base in the active site (Hu et al., 2015). We observed decreased prevalence of hydrogen bonds formed for these two residues in the ribo systems (Table 1).

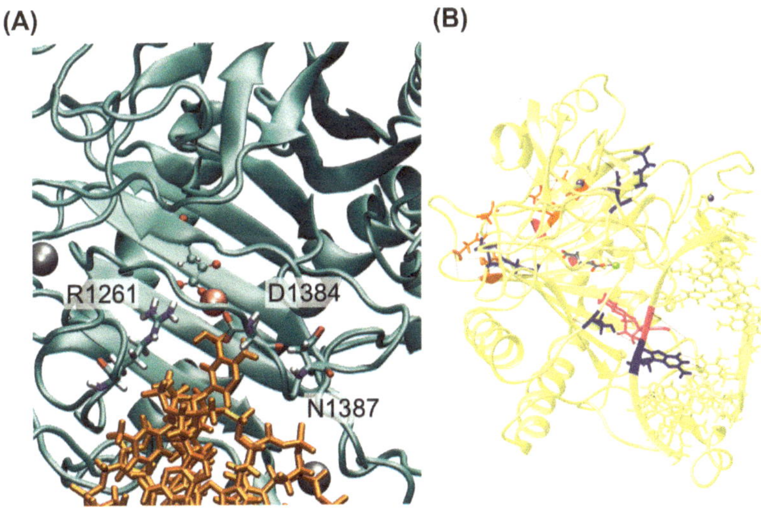

Fig. 11 (A) Residues surrounding the active site for the 5hmrC:5mdC, 5frC:5mdC, and 5carC:5mdC that are ≥ 20% different between the deoxy and ribo systems. (B) Residues with differences ≥ 20% in hydrogen bonding indicated in all systems with altered complementary strands (5mdC:dC, 5mdC:5mrC, 5mrC:5mrC:, and ss 5mrC). Blue residues form hydrogen bonds more prevalent in the reference system, orange residues form hydrogen bonds more prevalent in the systems with altered complementary strand, and pink residues have a system-dependent preference.

Table 1 Differences in hydrogen bonding between that exceed 20% of the total production time for the specified system.

Hydrogen acceptor	Hydrogen donor	Absolute difference (%)	System
ASN1387@OD1	5hmC Substrate	33.2950	5hmC and 5hmrC
ASP1384@OD2	ASN1387@ND2	29.3766	5fdC and 5frC
THR1393@OG1	ASN1387@ND2	24.8167	5fdC and 5frC
HIE1904@ND1	TRP1291@N	22.2400	5fdC and 5frC

Each of the systems with altered complementary strands (5mdC:dC, 5mdC:5mrC, 5mrC:5mrC:, and ss 5mrC) shared several residues with major differences in their hydrogen bonding (Fig. 11B). Each of the altered complementary strand systems can be compared with a reference system most closely matching the system with the unaltered nucleotide to identify the changes in hydrogen bonding patterns. Specifically, 5mdC:dC and 5mdC: 5mrC can be compared with 5mdC:5mdC; 5mrC:5mrC can be compared with 5mrC:5mdC; and ss 5mrC with ss 5mdC. The residues that showed hydrogen bonds with longer permanence throughout the production simulation in the reference systems correspond to residues that line the beta sheets enclosing the active site (N1266, E1268, D1376, Y1895, substrate, H1904, G7, T1259). Residues farther from the nucleic acid strand (N1403, T1399, D1412, N1403, E1178, R1176) showed longer lasting hydrogen bonds in the systems with altered complementary strands.

The hydrogen bond between residues E1165 and G1172 was observed for a longer time across simulations in the single stranded ribo system than in the ds altered complementary systems. This interaction takes place in a central area of the protein. The other major change in hydrogen bonding occurred between the substrate and H1904. The presence of that hydrogen bond was diminished by about 30% of the simulation time for 5mdC:5mrC, 5mrC:5mrC:, and ss 5mrC. However, for 5mdC:dC, the presence of the substrate - H1904 hydrogen bond increased by about 35%.

TET2 has been shown experimentally to have more oxidation events with 5mdC:dC than any other tested substrate, and that all-deoxy substrates were preferred to DNA-RNA duplexes or chimeras (DeNizio et al., 2019). Here, we identified several hydrogen bonds that change between these contexts to accommodate ribonucleotides in the active site. Our results indicate that residues distal from the nucleic acid strand form longer-lived (or new) hydrogen bonds when a ribonucleotide is present, while breaking hydrogen bonds in the area closest to the strand in order to allow for the additional ribo oxygen, which may provide one explanation for the experimental pattern of activity.

3.4 Energy decomposition analysis

There are several key residues that interact with α-KG (R1261, R1896, and S1898), impact iron chelation (H1382, D1384, and H1881), and recognize DNA (T1372, N1387, Y1902, and H1904) (Table 2). These residues have previously been identified, and some have been observed from mutations in patients with myeloid malignancies (Yin & Xu, 2016). Therefore, the inter–molecular interactions of these key residues are crucial, and changes in these interactions arising from differences in the substrate may have a significant impact in TET2 activity. The most pronounced changes in inter–molecular interactions in these residues include R1261, D1384, and R1896. R1261, for example, has a considerably smaller $E_{Tot}^{Non-bond}$ for the ds 5mC systems, while similarly favoring the ribo system.

An earlier investigation of TET2 mutations identified a loss of activity with the T1372E mutant, stalling at 5hmC, due to the disruption of an active site scaffold that could be recovered with a secondary Y1902A mutation (Liu et al., 2016). Because a similar stalling pattern arose with the ribo systems experimentally, we were interested in looking at commonalities between the 5hmdC T1372E mutant and 5mrC:5mdC. Consistent with our previous work, we found that the ribo systems did not exhibit any pronounced differences for the intermolecular interactions between the substrate and T1372 or Y1902 regardless of the identity or the position of the ribonucleotide modification (substrate or complement). However, the interaction energies between the substrate residue and the iron-coordinated aspartate (D1384) were much more comparable than any other systems studied here (Table 2). Accordingly, the substrate – D1384 interaction appears to be critical to TET2's function.

Expanding to the entire protein, differences between the deoxy and ribo systems were observed across each of the oxidation products (Fig. 12). Pronounced changes were seen in the ss systems, particularly in the two loops that interact with the DNA (residues 1256-1273, 1288-1312), as well as the three zinc-coordinating regions (Zn1: C1133, C1135, C1221; Zn2: C1193, C1271; Zn3: C1289, C1298, C1358, A1458, H1912) (Hu et al., 2013).

Approximately 39% of the residues between the 5caC systems showed differences ≥ 1 kcal mol^{-1}, with over 25% of the protein showing differences ≥ 10 kcal mol^{-1}. Thus, the ribo oxygen, in combination with the

Table 2 Inter–molecular interaction energy difference, $\Delta E_{Tot}^{Non-bond}$, for key residues between dC and rC systems.

	ss	ds			
	$\Delta E_{5mdC-5mrC}$	$\Delta E_{5hmdC-5hmrC}$	$\Delta E_{5fdC-5frC}$	$\Delta E_{5cadC-5carC}$	$\Delta E^{\star}_{5hmdC-5mrC}$
R1261†	51.78 ± 1.50	4.80 ± 1.10	15.15 ± 0.88	− 51.56 ± 2.15	11.92 ± 1.84
T1372	1.58 ± 0.15	0.47 ± 0.09	− 0.11 ± 0.07	− 8.11 ± 0.77	21.75 ± 0.06
H1382‡	1.23 ± 0.07	1.04 ± 0.21	0.78 ± 0.04	5.02 ± 0.13	1.94 ± 0.15
D1384‡	1.34 ± 1.18	− 13.75 ± 1.24	− 5.52 ± 0.56	45.11 ± 0.93	0.37 ± 2.16
N1387	− 2.67 ± 0.20	0.64 ± 0.22	− 1.23 ± 0.13	0.61 ± 0.26	− 0.54 ± 0.29
H1881‡	− 0.52 ± 0.05	− 0.28 ± 0.14	− 0.37 ± 0.08	3.62 ± 0.06	0.06 ± 0.08
R1896†	6.10 ± 0.07	− 0.32 ± 0.09	− 0.03 ± 0.05	− 21.76 ± 0.27	− 0.01 ± 0.11
S1898†	− 0.08 ± 0.02	0.20 ± 0.02	0.01 ± 0.01	0.13 ± 0.02	− 0.33 ± 0.02
Y1902	3.51 ± 0.18	0.87 ± 0.22	− 0.31 ± 0.14	1.57 ± 0.28	− 0.52 ± 0.26
H1904	1.36 ± 0.29	1.63 ± 0.21	0.29 ± 0.51	− 0.21 ± 0.10	1.75 ± 0.94

†, denotes a residue that interacts with α-KG; ‡, signifies a residue that coordinates with Fe(II), and no symbol means that the residue is involved in DNA recognition; $\Delta E_{5hmdC-5mrC}$ column denotes that reference 5hmdC system corresponds to the T1372E mutation. All values are the average of each system ± avg. st. dev. in kcal mol^{-1}.

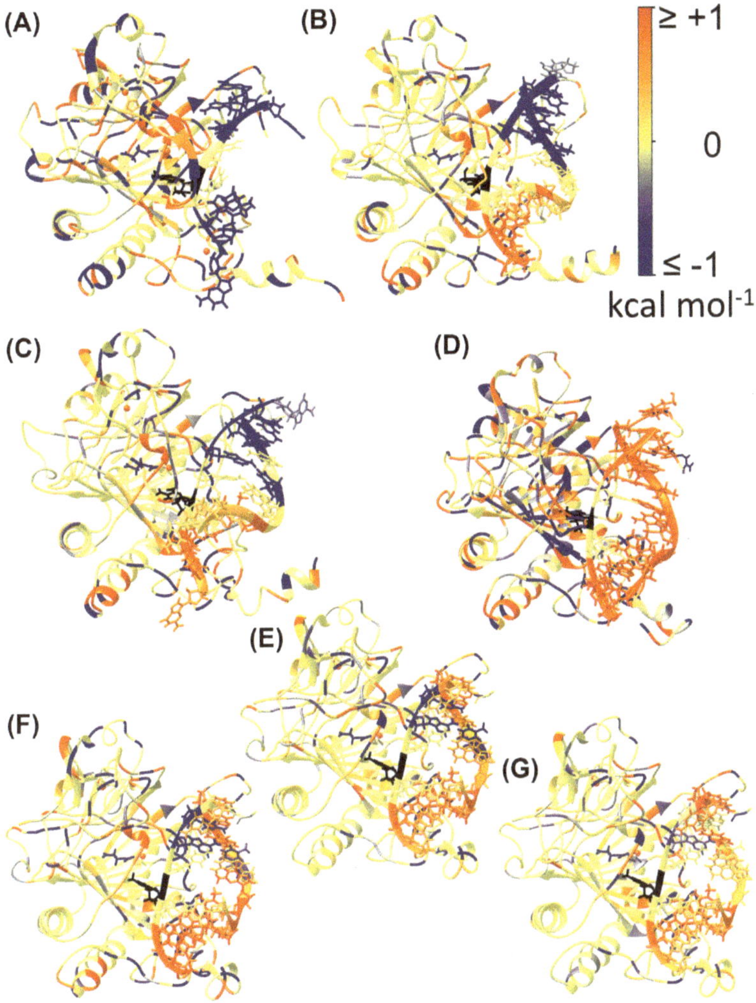

Fig. 12 Differences between the total non—bonded interactions (Coulomb and Van der Waals) mapped onto the TET2 structure for (A) ss 5mdC - ss 5mrC, (B) 5hmdC:5mdC - 5hmrC:5mdC, (C) 5fdC:5mdC - 5frC:5mdC, (D) 5cadC:5mdC - 5carC:5mdC, (E) 5mdC:5mdC - 5mdC:5mrC (F) 5mdC:5mdC - 5mrC:5mrC, and (G) 5mdC:5mdC - 5mdC:dC. Residues colored blue are favored in the first system and residues colored orange are favored in the second system. The target nucleotide is shown in black.

two oxygens inherent to 5caC, adds additional obstructive interactions to the already crowded active site.

Comparing the systems with changes to the complementary strands, two additional residues (excluding the active site, DNA-interacting loops, and zinc-coordinating regions) had large ΔE values. For $\Delta E_{5mdC:5mdC-5mrC:5mrC}$, these included H1386 $(-3.55 \pm 0.50\ \text{kcal mol}^{-1})$ and K1462 $(8.11 \pm 0.58\ \text{kcal mol}^{-1})$. For $\Delta E_{5mdC:5mdC-5mdC:dC}$, this again included K1462 $(4.05 \pm 0.21\ \text{kcal mol}^{-1})$. Both H1386 and K1462 are located in non—structured regions on the same face of the protein as the DNA helix, and are close enough to the DNA strand that their inherent flexibility may impact the magnitude of their interaction (Fig. 13).

Similar to the hydrogen bonding analysis results, the addition of oxygen in the ribonucleotide induced disruptions to the non—bonded interactions, giving rise to instability in the active site. These differences were seen predominantly in the active site and nucleic acid backbone, with farther-reaching effects for the ss systems. Change from 5mdC to dC in the complementary strand better localized these energetic differences to the backbone. Together, both the changes and hydrogen bonding and EDA agree with the experimental preference for ds, deoxy substrates, due to the active site instability of ribonucleotides and, to a lesser extent, the flexibility of ssDNA.

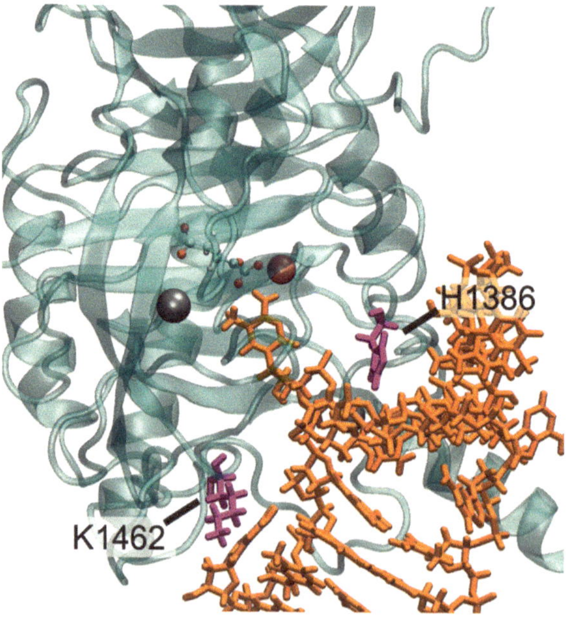

Fig. 13 The H1384 and K1462 residues, located in unstructured loops, that exhibit large changes in their total non—bonded interactions for comparisons between the 5mdC: 5mdC system and the systems with altered complementary strands.

4. Conclusions

We have performed a computational investigation of the effect of the inclusion of various modified ribonucleotides at the substrate and complement positions on the structure and dynamics of TET2. Our results indicate that replacing a modified cytosine deoxynucleotide by a modified cytosine ribonucleotide in the target position of a TET2-DNA complex has a large and fairly consistent impact on the backbone structure, energetics, and dynamics of the system. Our results suggest that ribonucleotide substrates may be disfavored due to their instability in the active site, brought about by repulsive interactions with the additional oxygen moiety on the ribose ring, and a significant shifting of the substrate structure from a B-form DNA helix into something sharing characteristics with A-form, B-form, and "F-form" DNA. Interestingly, having a dC nucleobase in the complementary strand can help stabilize ribonucleotide substrates, while a ribonucleotide in the complementary strand has a similar overall effect as the inclusion of a ribonucleotide target at the substrate position. Thus, it appears that the composition of both the substrate are important for TET2's function, and that deviations from typical DNA substrates further complicate the complete oxidation of 5mC.

Acknowledgments

This work was funded by NIH via R01GM108583 and R01GM118501. Computing time from CASCaM with partial support from NSF Grant No. CHE−1531468 is acknowledged. The authors thank Dr. A.R. Walker for assistance with the parametrization of the RNA cytosine-derivatives.

References

Aik, W., McDonough, M. A., Thalhammer, A., Chowdhury, R., & Schofield, C. J. (2012). Role of the jelly-roll fold in substrate binding by 2-oxoglutarate oxygenases. *Current Opinion in Structural Biology, 22*(6), 691−700. http://dx.doi.org/10.1016/j.sbi.2012.10.001.

Babcock, M. S., Pednault, E. P., & Olson, W. K. (1994). Nucleic acid structure analysis: Mathematics for local cartesian and helical structure parameters that are truly comparable between structures. *Journal of Molecular Biology, 237*(1), 125−156. http://dx.doi.org/10.1006/jmbi.1994.1213.

Bayly, C. I., Cieplak, P., Cornell, W., & Kollman, P. A. (1993). A well-behaved electrostatic potential based method using charge restraints for deriving atomic charges: The RESP model. *The Journal of Physical Chemistry, 97*(40), 10269−10280. http://dx.doi.org/10.1021/j100142a004.

Berendsen, H. J. C., Postma, J. P. M., van Gunsteren, W. F., DiNola, A., & Haak, J. R. (1984). Molecular dynamics with coupling to an external bath. *The Journal of Chemical Physics, 81*(8), 3684−3690. http://dx.doi.org/10.1063/1.448118.

Bird, A. (2002). DNA methylation patterns and epigenetic memory. *Genes & Development, 16*(1), 6–21. http://dx.doi.org/10.1101/gad.947102.

Brooks, S. C., Adhikary, S., Rubinson, E. H., & Eichman, B. F. (2013). Recent advances in the structural mechanisms of DNA glycosylases. *Biochimica et Biophysica Acta, 1834*(1), 247–271. http://dx.doi.org/10.1016/j.bbapap.2012.10.005 (Special Section: Chemistry and mechanism of phosphatases, diesterases and triesterases).

Case, D., Cerutti, D., Cheatham, T., III, Darden, T., Duke, R., Giese, T., et al. (2017). *Amber16* (San Francisco).

Chen-Kiang, S., Nevins, J. R., & Darnell, J. E. (1979). N-6-methyl-adenosine in adenovirus type 2 nuclear RNA is conserved in the formation of messenger RNA. *Journal of Molecular Biology, 135*(3), 733–752. http://dx.doi.org/10.1016/0022-2836(79)90174-8.

DeNizio, J. E., Liu, M. Y., Leddin, E. M., Cisneros, G. A., & Kohli, R. M. (2019). Selectivity and promiscuity in TET-mediated oxidation of 5-methylcytosine in DNA and RNA. *Biochemistry, 58*(5), 411–421. http://dx.doi.org/10.1021/acs.biochem.8b00912 (PMID: 30387995).

Dewage, S. W., & Cisneros, G. A. (2015). Computational analysis of ammonia transfer along two intramolecular tunnels in *Staphylococcus aureus* glutamine-dependent amidotransferase (GatCAB). *The Journal of Physical Chemistry B, 119*(9), 3669–3677. http://dx.doi.org/10.1021/jp5123568 (PMID: 25654336).

Dowle, M., & Srinivasan, A. (2018). *data.table: Extension of 'data.frame' [Computer software manual].* Retrieved from https://CRAN.R-project.org/package=data.table (R package version 1.11.8.

Drohat, A. C., & Coey, C. T. (2016). Role of base excision "repair" enzymes in erasing epigenetic marks from DNA. *Chemical Reviews, 116*(20), 12711–12729. http://dx.doi.org/10.1021/acs.chemrev.6b00191.

Dupradeau, F.-Y., Pigache, A., Zaffran, T., Savineau, C., Lelong, R., Grivel, N., et al. (2010). The R.E.D. tools: Advances in RESP and ESP charge derivation and force field library building. *Physical Chemistry Chemical Physics, 12*, 7821–7839. http://dx.doi.org/10.1039/C0CP00111B.

Elias, A. A., & Cisneros, G. A. (2014). Chapter two - computational study of putative residues involved in DNA synthesis fidelity checking in Thermus aquaticus DNA polymerase I. In T. Karabencheva-Christova (Ed.), *Biomolecular modelling and simulations* (Vol. 96, pp. 39–75). Academic Press. http://dx.doi.org/10.1016/bs.apcsb.2014.06.003.

Essmann, U., Perera, L., Berkowitz, M. L., Darden, T., Lee, H., & Pedersen, L. G. (1995). A smooth particle mesh Ewald method. *The Journal of Chemical Physics, 103*(19), 8577–8593. http://dx.doi.org/10.1063/1.470117.

Fang, D., & Cisneros, G. A. (2014). Alternative pathway for the reaction catalyzed by DNA dealkylase AlkB from ab initio QM/MM calculations. *Journal of Chemical Theory and Computation, 10*(11), 5136–5148.

Fang, D., Lord, R. L., & Cisneros, G. A. (2013). Ab initio QM/MM calculations show an intersystem crossing in the hydrogen abstraction step in dealkylation catalyzed by AlkB. *The Journal of Physical Chemistry B, 117*(21), 6410–6420. http://dx.doi.org/10.1021/jp403116e (PMID: 23642148).

Fu, L., Guerrero, C. R., Zhong, N., Amato, N. J., Liu, Y., Liu, S., et al. (2014). Tet-mediated formation of 5-hydroxymethylcytosine in RNA. *Journal of the American Chemical Society, 136*(33), 11582–11585. http://dx.doi.org/10.1021/ja505305z (PMID: 25073028).

Götz, A. W., Williamson, M. J., Xu, D., Poole, D., Le Grand, S., & Walker, R. C. (2012). Routine microsecond molecular dynamics simulations with AMBER on GPUs. 1. Generalized Born. *Journal of Chemical Theory and Computation, 8*(5), 1542–1555. http://dx.doi.org/10.1021/ct200909j (PMID: 22582031).

Graham, S. E., Syeda, F., & Cisneros, G. A. (2012). Computational prediction of residues involved in fidelity checking for DNA synthesis in DNA polymerase I. *Biochemistry, 51*(12), 2569–2578. http://dx.doi.org/10.1021/bi201856m (PMID: 22397306).

Hardwick, J. S., Ptchelkine, D., El-Sagheer, A. H., Tear, I., Singleton, D., Phillips, S. E. V., Lane, A. N., & Brown, T. (2017). 5-Formylcytosine does not change the global structure of DNA. *Nature Structural & Molecular Biology, 24*(6), 544–552.

Hong, S., & Cheng, X. (2016). DNA base flipping: A general mechanism for writing, reading, and erasing DNA modifications. In A. Jeltsch, & R. Z. Jurkowska (Eds.), *DNA methyltransferases - role and function* (pp. 321–341). Cham: Springer International Publishing. http://dx.doi.org/10.1007/978-3-319-43624-1_14.

Hu, L., Li, Z., Cheng, J., Rao, Q., Gong, W., Liu, M., et al. (2013). Crystal structure of TET2-DNA complex: Insight into TET-mediated 5mC oxidation. *Cell, 155*(7), 1545–1555. http://dx.doi.org/10.1016/j.cell.2013.11.020.

Hu, L., Lu, J., Cheng, J., Rao, Q., Li, Z., Hou, H., et al. (2015). Structural insight into substrate preference for TET-mediated oxidation. *Nature, 527*, 118–122. http://dx.doi.org/10.1038/nature15713.

Humphrey, W., Dalke, A., & Schulten, K. (1996). VMD: Visual molecular dynamics. *Journal of Molecular Graphics, 14*(1), 33–38. http://dx.doi.org/10.1016/0263-7855(96)00018-5.

Jin, B., Li, Y., & Robertson, K. D. (2011). DNA methylation: Superior or subordinate in the epigenetic hierarchy? *Genes & Cancer, 2*(6), 607–617. http://dx.doi.org/10.1177/1947601910393957.

Jones, P. A. (2012). Functions of DNA methylation: Islands, start sites, gene bodies and beyond. *Nature Reviews Genetics, 13*, 484–492. http://dx.doi.org/10.1038/nrg3230.

Jorgensen, W. L., Chandrasekhar, J., Madura, J. D., Impey, R. W., & Klein, M. L. (1983). Comparison of simple potential functions for simulating liquid water. *The Journal of Chemical Physics, 79*(2), 926–935. http://dx.doi.org/10.1063/1.445869.

Kizaki, S., & Sugiyama, H. (2014). CGmCGCG is a versatile substrate with which to evaluate Tet protein activity. *Organic and Biomolecular Chemistry, 12*, 104–107. http://dx.doi.org/10.1039/C3OB41823E.

Ko, M., An, J., Bandukwala, H. S., Chavez, L., Äijö, T., Pastor, W. A., et al. (2013). Modulation of TET2 expression and 5-methylcytosine oxidation by the CXXC domain protein IDAX. *Nature, 497*, 122–126. http://dx.doi.org/10.1038/nature12052.

Kohli, R. M., & Zhang, Y. (2013). TET enzymes, TDG and the dynamics of DNA demethylation. *Nature, 502*, 472–479. http://dx.doi.org/10.1038/nature12750.

Liu, M. Y., Torabifard, H., Crawford, D. J., DeNizio, J. E., Cao, X.-J., Garcia, B. A., et al. (2016). Mutations along a TET2 active site scaffold stall oxidation at 5-hydroxymethylcytosine. *Nature Chemical Biology, 13*, 181–187.

Lorsbach, R. B., Moore, J., Mathew, S., Raimondi, S. C., Mukatira, S. T., & Downing, J. R. (2003). TET1, a member of a novel protein family, is fused to MLL in acute myeloid leukemia containing the t(10;11)(q22;q23). *Leukemia, 17*, 637–641. http://dx.doi.org/10.1038/sj.leu.2402834.

Olson, W. K., Bansal, M., Burley, S. K., Dickerson, R. E., Gerstein, M., Harvey, S. C., et al. (2001). A standard reference frame for the description of nucleic acid base-pair geometry. *Journal of Molecular Biology, 313*(1), 229–237. http://dx.doi.org/10.1006/jmbi.2001.4987.

Packer, M. J., & Hunter, C. A. (1998). Sequence-dependent DNA structure: The role of the sugar-phosphate backbone. *Journal of Molecular Biology, 280*(3), 407–420. http://dx.doi.org/10.1006/jmbi.1998.1865.

Pettersen, E. F., Goddard, T. D., Huang, C. C., Couch, G. S., Greenblatt, D. M., Meng, E. C., et al. (2004). UCSF Chimera—a visualization system for exploratory research and analysis. *Journal of Computational Chemistry, 25*(13), 1605–1612. http://dx.doi.org/10.1002/jcc.20084.

Plate, T., & Heiberger, R. (2016). *abind: Combine multidimensional arrays [Computer software manual]*. Retrieved from https://CRAN.R-project.org/package=abind (R package version 1.4-5).

R Core Team. (2018). *R: A language and environment for statistical computing [Computer software manual]*. Vienna, Austria. Retrieved from https://www.R-project.org/.

Raiber, E.-A., Murat, P., Chirgadze, D. Y., Beraldi, D., Luisi, B. F., & Balasubramanian, S. (2015). 5-formylcytosine alters the structure of the DNA double helix. *Nature Structural & Molecular Biology, 22*, 44–49. http://dx.doi.org/10.1038/nsmb.2936.

Rasmussen, K. D., & Helin, K. (2016). Role of TET enzymes in DNA methylation, development, and cancer. *Genes & Development, 30*(7), 733–750. http://dx.doi.org/10.1101/gad.276568.115.

Roe, D. R., & Cheatham, T. E. (2013). PTRAJ and CPPTRAJ: Software for processing and analysis of molecular dynamics trajectory data. *Journal of Chemical Theory and Computation, 9*(7), 3084–3095. http://dx.doi.org/10.1021/ct400341p (PMID: 26583988).

Šali, A., & Blundell, T. L. (1993). Comparative protein modelling by satisfaction of spatial restraints. *Journal of Molecular Biology, 234*(3), 779–815. http://dx.doi.org/10.1006/jmbi.1993.1626.

Salomon-Ferrer, R., Götz, A. W., Poole, D., Le Grand, S., & Walker, R. C. (2013). Routine microsecond molecular dynamics simulations with AMBER on GPUs. 2. Explicit solvent particle mesh Ewald. *Journal of Chemical Theory and Computation, 9*(9), 3878–3888. http://dx.doi.org/10.1021/ct400314y (PMID: 26592383).

Schafmeister, C. E. A. F., Ross, W. S., & Romanovski, V. (1995). *LEAP*. San Francisco: University of California.

Shen, L., Song, C.-X., He, C., & Zhang, Y. (2014). Mechanism and function of oxidative reversal of DNA and RNA methylation. *Annual Review of Biochemistry, 83*(1), 585–614. http://dx.doi.org/10.1146/annurev-biochem-060713-035513.

Thalhammer, A., Hansen, A. S., El-Sagheer, A. H., Brown, T., & Schofield, C. J. (2011). Hydroxylation of methylated CpG dinucleotides reverses stabilisation of DNA duplexes by cytosine 5-methylation. *Chemical Communications, 47*, 5325–5327. http://dx.doi.org/10.1039/C0CC05671E.

Torabifard, H., & Cisneros, G. A. (2018). Insight into wild-type and t1372e TET2-mediated 5hmc oxidation using ab initio QM/MM calculations. *Chemical Science, 9*, 8433–8445. http://dx.doi.org/10.1039/c8sc02961j.

Tsai, C.-L., & Tainer, J. A. (2013). Probing DNA by 2-OG-dependent dioxygenase. *Cell, 155*(7), 1448–1450. http://dx.doi.org/10.1016/j.cell.2013.12.002.

Vanquelef, E., Simon, S., Marquant, G., Garcia, E., Klimerak, G., Delepine, J. C., et al. (2011, 05). R.E.D. Server: A web service for deriving RESP and ESP charges and building force field libraries for new molecules and molecular fragments. *Nucleic Acids Research, 39*(Suppl. l_2), W511–W517. http://dx.doi.org/10.1093/nar/gkr288.

Wang, F., Becker, J.-P., Cieplack, P., & Dupradeau, F.-Y. (2013). *R.E.D. python: Object oriented programming for AMBER force fields*. Université de Picardie-Jules Verne, Sanford Burnham Prebys Medical Discovery Institute.

Wickham, H. (2017). *tidyverse: Easily install and load the 'tidyverse' [Computer software manual]*. Retrieved from https://CRAN.R-project.org/package=tidyverse (R package version 1.2.1.

Williams, K., Christensen, J., & Helin, K. (2012). DNA methylation: TET proteins—guardians of CpG islands? *EMBO Reports, 13*(1), 28–35. http://dx.doi.org/10.1038/embor.2011.233.

Yin, X., & Xu, Y. (2016). Structure and function of TET enzymes. In A. Jeltsch, & R. Z. Jurkowska (Eds.), *DNA methyltransferases - role and function* (pp. 275–302). Cham: Springer International Publishing. http://dx.doi.org/10.1007/978-3-319-43624-1_12.

CHAPTER FOUR

Structure-function relationships in KDM7 histone demethylases

Shobhit S. Chaturvedi[1], Rajeev Ramanan[1], Sodiq O. Waheed, Tatyana G. Karabencheva-Christova* and Christo Z. Christov*

Department of Chemistry, Michigan Technological University, Houghton, MI, United States

*Corresponding authors: E-mail: tatyanak@mtu.edu, christov@mtu.edu

Contents

1. Introduction 114
2. Biological and pathological role 115
3. Structure and mechanism 115
4. Dynamics of KDM7 enzyme-substrate complex 117
5. Reactivity studies of model system 121
6. Inhibitors of KDM7 122
7. Conclusions 123
References 124

Abstract

The demethylation of lysine residues of histone proteins is a key epigenetic mechanism in cells. The enzymes that catalyze these processes are called histone demethylases (KDMs). The largest family of KDMs is the Jumonji C (JmjC) domain-containing enzymes; these includes KDM2-7 subfamily of enzymes. The JmjC proteins are Fe(II) and 2-Oxoglutarate (2OG) - dependent dioxygenases that couple substrate oxidation to decarboxylation of 2OG to form succinate and CO_2. The KDM7 subfamily of enzymes - PHF8 (KDM7B) and KIAA1718 (KDM7A) are human JmjC 2OG-dependent N^{ε}-methyl lysine demethylases and are involved in demethylation of lysine residues in histones such as H3K27me2/1, H3K9me2/1 and H4K20me1. These enzymes are involved in multiple pathologic processes, including cancers and mental retardation. In this chapter, we present the current state of the art in the structural, biochemical and computational studies of KDM7 enzymes.

[1] SSC and RR have equal contributions.

Advances in Protein Chemistry and Structural Biology, Volume 117
ISSN 1876-1623
https://doi.org/10.1016/bs.apcsb.2019.08.005

© 2019 Elsevier Inc.
All rights reserved.

1. Introduction

Post-translational modifications (PTM) to histone proteins, in the nucleosome, are known to play a crucial role in gene regulation and expression. The nucleosome contains DNA wrapped around an octamer (Luger & Hansen, 2005) built of two molecules from the following histone proteins - H2A, H2B, H3 and H4. The histone octamer and DNA in the nucleosome are linked via Histone H1 (Fig. 1). An important PTM to histone proteins is the methylation of lysine or arginine residues on unstructured N-terminal tails protruding out of histones in nucleosomes. The methylation state on lysine residues can be mono-, di- and trimethylation (me1, me2, and me3) (Bannister & Kouzarides, 2004). Based on the lysine methylation state and the surrounding amino acid sequence, the reader domains in proteins, e.g. the PHD domain, can target methylated lysines. The state and position of the N^{ε}-lysine methylation determine its role as an activator or repressor of the transcription, e.g. methylation on H3K4, H3K36 and H3K79 activates, while H3K9, H3K27 and H4K20 methylation represses the transcription. Demethylation of lysine N^{ε}-methylation group is performed by a group of enzymes called histone lysine demethylases (KDMs).

The histone lysine demethylase (KDM) family is divided into two subfamilies. The KDM1 subfamily, also known as lysine-specific demethylase (LSD) subfamily, which are flavin-dependent enzymes closely related to monoamine oxidases (MAOs) (Shi et al., 2004). The second subfamily is the Jumonji C (JmjC) domain-containing enzymes; these include KDM2-7 subfamily of enzymes. The JmjC proteins are Fe(II)- and 2-Oxoglutarate (2OG) - dependent dioxygenases, that couple substrate oxidation to

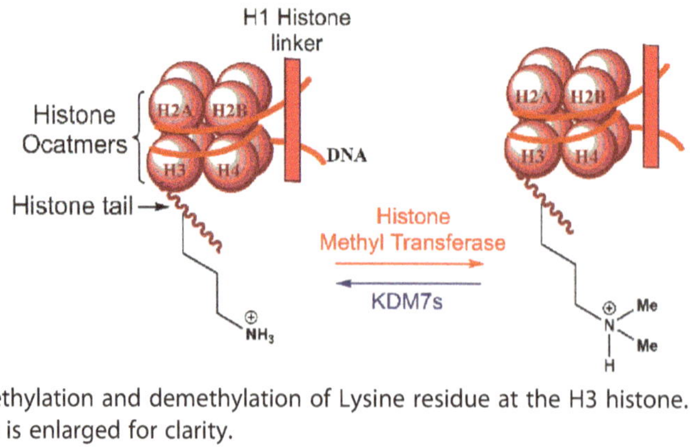

Fig. 1 Methylation and demethylation of Lysine residue at the H3 histone. The tail of the lysine is enlarged for clarity.

decarboxylation of 2OG, forming succinate and CO_2 (Hausinger, 2004; Walport, Hopkinson, & Schofield, 2012). KDM7 subfamily of enzymes - PHF8 (KDM7B) and KIAA1718 (KDM7A) are human JmjC 2OG-dependent, N^{ε}-methyl lysine demethylases that are involved in demethylation of H3K27me2/1, H3K9me2/1 and H4K20me1 (Björkman et al., 2012; Horton et al., 2010). The KDM7 subfamily also includes KDM7C enzyme but only *in vivo* activity is reported for KDM7C.

2. Biological and pathological role

The KDM7 subfamily plays a major role in cancer. KDM7 can promote the growth of certain tumor cells while, at the same time, it can also inhibit some other types tumors. Specifically, KDM7A enzymes effectively hinder xenograft tumor growth, by regulating angiogenesis, while the removal of KDM7B in cancer cells are effective in inhibiting the growth and invasion of prostate cancer (Björkman et al., 2012; Ma et al., 2015). Other functional roles of KDM7B include overexpression in laryngeal and hypopharyngeal cancer (Zhu et al., 2015), breast cancer (Wang et al., 2016), esophageal cancer (Sun et al., 2013) and non-small cell lung cancer (Shen, Pan, & Zhao, 2014). Mutations in KDM7B can lead to fatal biological dysfunction. It is reported that the mutation of Phe279 in KDM7B for Ser initiates the inherited X-linked mental retardation (Abidi, Miano, Murray, & Schwartz, 2007; Koivisto et al., 2007; Laumonnier et al., 2005; Loenarz et al., 2009). However, reports suggest an increased retinoic acid response in acute promyelocytic leukemia in the presence of KDM7B (Arteaga et al., 2013). While the presence of KDM7A and KDM7B largely promote growth and invasion of cancer cells, KDM7C majorly suppress tumors. The tumor cells that are suppressed by KDM7C include breast cancer (Sinha et al., 2008), head and neck squamous cell carcinoma (Ghosh et al., 2012), as well as colon and stomach cancers (Lee et al., 2015).

3. Structure and mechanism

KDM7B (PHF8) has a reader domain called Plant Homeodomain (PHD) at the N-terminal and a catalytic domain called Jumonji C domain (JmjC) at the C-terminal (Fig. 2). A flexible linker region connects the PHD and the JmjC domains. The PHD domain is responsible for the initial binding of the histone H3 substrate with K4me3, and then with the help of

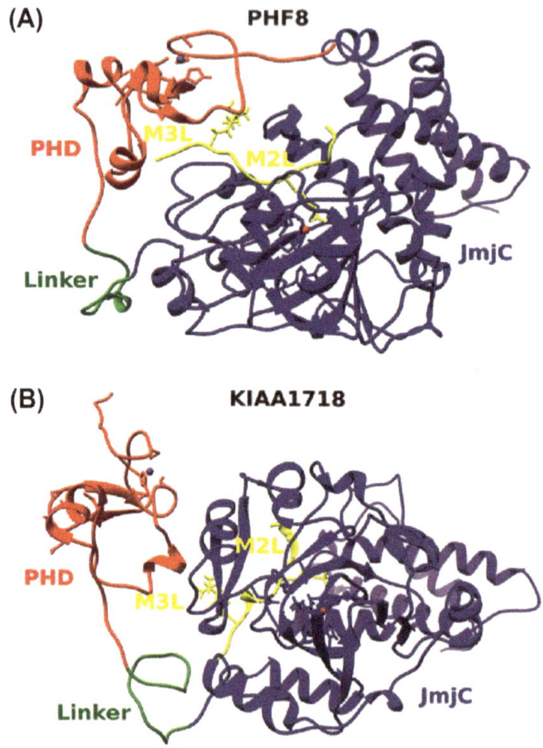

Fig. 2 Views from MD structures of (A) PHF8 and (B) KIAA1718. Averaged structures from clustering are presented. *Copied with permission from Chaturvedi, S. S., Ramanan, R., Waheed, S. O., Ainsley, J., Evison, M., Ames, J. M., et al. (2019). Conformational dynamics underlies different functions of human KDM7 histone demethylases.* Chemistry - A European Journal, *25(21), 5422–5426.*

the flexible linker, the catalytic domain JmjC binds at K9me2 for demethylation reaction. The JmjC domain has a conserved double-stranded β-helix (DSBH), which has four anti-parallel beta sheets spanning over the active site (Aik, McDonough, Thalhammer, Chowdhury, & Schofield, 2012). The active site is located inside the DSBH and contains Fe, 2OG co-substrate, and three conserved amino acids - His-Asp-His triad from the protein (Hausinger, 2004; Walport et al., 2012). The 2OG co-substrate chelates to Fe using the oxygens of the C1 carboxylate and C2 keto group, while the other three coordination position at the Fe are occupied by His-Asp-His triad.

KDM7A (KIAA1718) has a structure similar to KDM7B with a PHD domain at the N-terminal, and JmjC at the C-terminal (Fig. 2). The difference between the KDM7A and KDM7B is mainly attributed to the difference in flexibility and the length of their linker region (Horton et al., 2010; Yue et al., 2010). Similar to KDM7B, the PHD domain of KDM7A binds to H3K4me3 but due to a rigid linker region, is unable to position the catalytic JmjC domain to nearby H3K9me2 for demethyation catalysis. The active site is very similar to KDM7B and is located inside DSBH and constitute a Fe co-factor, 2OG co-substrate, and three conserved amino acids, the His-Asp-His triad from the protein (Horton et al., 2010; Yue et al., 2010).

KDM7C shows little demethylase activity and is built from the same domains. Similar to other KDM7 enzymes, the PHD domain of KDM7C also binds at H3K4me3. The catalytic JmjC domain has a similar DSBH, and the coordination sphere of Fe is also similar to the JmjC domains of KDM7B and 7A (Horton, Upadhyay, Hashimoto, Zhang, & Cheng, 2011) with the only difference being the fifth ligand is a tyrosine (Tyr321). Presence of a sterically bulky Tyr321 affects the functions of the enzyme. Oxygen binding in KDM7C enzymes is not as effective due to a slight displacement of Fe by the Tyr321.

The KDM7 subfamily is expected to show a similar reaction mechanism like other 2OG dependent enzymes. The reactive Fe-center is six-coordinated containing 2OG, the substrate and dioxygen molecule. The proposed reaction mechanism for KDM7 begins by binding of O_2 to Fe center at the position of a water molecule (Walport et al., 2012; Hausinger, 2004). The formed ferric-superoxo radical complex ($Fe(III)$-$OO^{\cdot-}$) attacks the C2 carbon atom of 2OG leading to decarboxylation and formation of succinate and the ferryl-oxo species (Scheme 1). The reactive ferryl-oxo Fe(IV) =O species formed then abstracts a hydrogen atom from the lysine Nε-methyl group and forms a ferric-hydroxyl Fe(III)-OH species which performs rebound hydroxylation to form hydroxylated substrate as a product (Hausinger, 2004; Walport et al., 2012) which further non-enzymatically converts to carbinolamine that leaves the active site and break down to formaldehyde and demethylated lysine substrate (Hausinger, 2004; Walport et al., 2012).

4. Dynamics of KDM7 enzyme-substrate complex

One microsecond MD simulation of KDM7B and KDM7A in complex with Fe, 2OG, and $H3_{1\text{-}14}$K4me3-K9me2 substrate shows the effect of conformational dynamics on the enzyme-substrate complex (Chaturvedi et al., 2019). The metal non-binding C5 carboxylate of 2OG in KDM7B

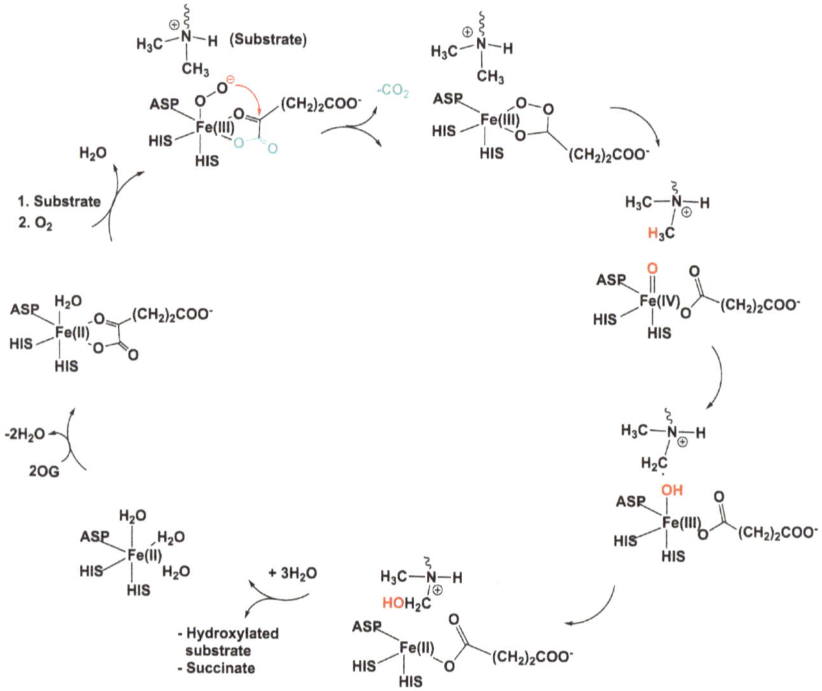

Scheme 1 Proposed reaction scheme for KDM7 enzymes.

interacts with the side chain of K264 (which is a feature of JmjC family of enzymes), Y257 and T244 (Fig. 3). Similarly, the C5 carboxylate of 2OG in KDM7A forms a hydrogen bond with K299, Y292, and T279 (Chaturvedi et al., 2019) (Fig. 3). In other KDMs except for KDM2, the tyrosine residue is replaced by an asparagine residue. A threonine (T244/T279) residue is also present in the C5 carboxylate binding pocket of KDM7B, KDM7A similar to KDM2A and KDM6A but phenylalanine is present at this position in KDM4 and KDM5 sub-families.

The MD simulation of KDM7B showed that the $H3_{1-14}K4me3$-K9me2 substrate could adopt an extended conformation to as observed in the crystal structure (Chaturvedi et al., 2019). Both the MD study and the crystal structures showed S192 and I191 as the two important residues which stabilize the substrate (H3K9me2) orientation at the KDM7B active site (Chaturvedi et al., 2019; Yue et al., 2010). Out of the two methyl groups of H3K9me2, one interacts with hydrophobic sidechains of Y234 and L236 while the other methyl group points toward Fe-coordinated D249 and the sidechain of N333. The rotation between the methyl groups is restricted due to the

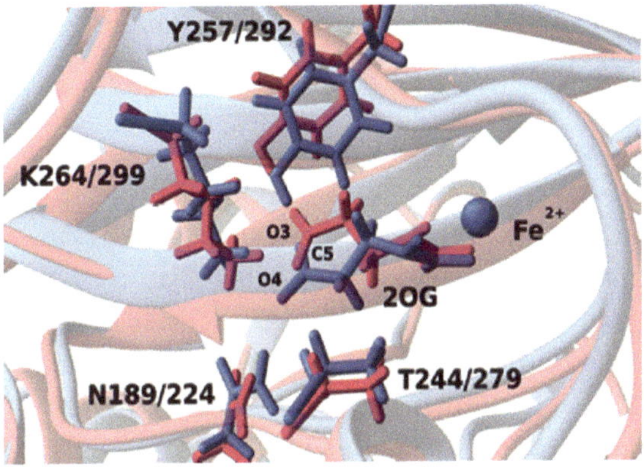

Fig. 3 Interactions of 2OG in the KDM7B (Pink) and KDM7A (Blue) active sites. Residues labeled as 1-letter code KDM7B/KDM7A. *Copied with permission from Chaturvedi, S. S., Ramanan, R., Waheed, S. O., Ainsley, J., Evison, M., Ames, J. M., et al. (2019). Conformational dynamics underlies different functions of human KDM7 histone demethylases.* Chemistry - A European Journal, *25(21), 5422–5426.*

hydrogen bond between NZ of H3K9me2 and O1 of 2OG. Importantly the methyl groups stay closer to the Fe, which is crucial for catalysis. The H3K4me3 side chain of the substrate makes extensive cation-π interactions with Y7, Y14, and W29 residues of the PHD domain (Chaturvedi et al., 2019).

In terms of domain flexibility, the PHD domain in both KDM7B and KDM7A showed higher flexibility as compared to the JmjC domain. The DSBH fold in the JmjC domain showed higher stability with a very low RMSD of 1.3 Å and 0.6 Å for KDM7B and KDM7A, respectively (Chaturvedi et al., 2019). The linker region in KDM7B is more flexible than that of KDM7A. It was suggested that the increased flexibility of KDM7B linker has a role in selective demethylation of H3K9me2 (Chaturvedi et al., 2019). In the MD study, the KDM7B linker showed a decreasing trend in a radius of gyration, indicating a bent conformation between the two domains PHD and JmjC in KDM7B. The bent conformation between the reader and the catalytic domain makes it possible for the PHD domain to bind at H3K4me3 and the catalytic domain to bind at a nearby H3K9me2 for selective demethylation (Chaturvedi et al., 2019). While in KDM7A, the linker stays in a more extended conformation, and the larger separation between both domains does not allow the PHD domain to bind at H3K4me3 and productive catalysis to take place at the H3K9me2. Role of the collective correlated motions using Dynamic Cross-Correlation

Analysis (DCCA) in KDM7B and KDM7A, for the effective binding of substrates was explored. KDM7B showed positive correlations within both the PHD domain and JmjC domains, while they showed negative (anti) correlations with respect to each other. The negative correlated motions between the domains help in independent and close interactions of the domains with the substrate (Chaturvedi et al., 2019). KDM7A showed more intense collective motions than in KDM7B (Chaturvedi et al., 2019) since KDM7A possess a more rigid linker, which results in better coupling of motions between the two domains. In silico modified KDM7A with KDM7B linker KDM7A* (KIAA1718*) showed more flexibility than the natural enzyme. Unlike KDM7A, the new KDM7A* prefers a bent conformation between the PHD and JmjC domain and hence a better binding at H3K9me2 for demethylation (Chaturvedi et al., 2019).

The effect of conformational flexibility of protein environment on the geometric and the electronic structure of the active site was explored using QM/MM optimization of KDM7B MD snapshots in complex with Fe, 2OG, and $H3_{1-14}$K4me3-K9me2 substrate. The coordination of the Fe was maintained, and it showed +2 oxidation state based on spin density which agrees with CD/MCD experiments on similar 2OG enzymes. The QM/MM optimized geometry showed similar interactions for 2OG and substrate, as seen in MD simulations (Chaturvedi et al., 2019). The important difference observed in QM/MM optimization was in the orientation of two methyl groups of substrates toward the Fe-center (Fig. 4). The five

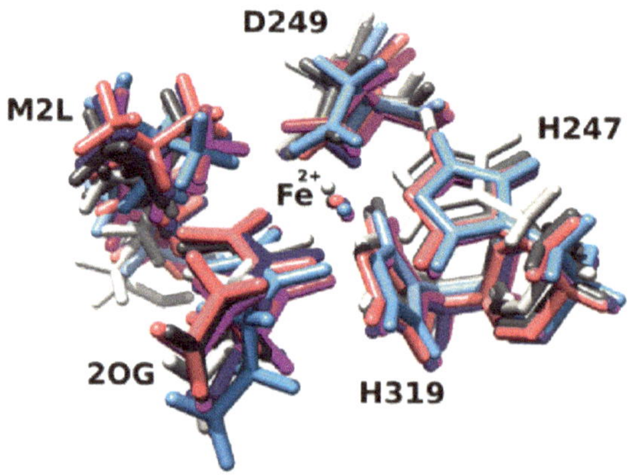

Fig. 4 Flexibility and movement at the active site of KDM7B reproduced by superimposing multiple QM/MM optimized geometries. H3K9me2 is labeled as M2L. *Copied with permission from Chaturvedi, S. S., Ramanan, R., Waheed, S. O., Ainsley, J., Evison, M., Ames, J. M., et al. (2019). Conformational dynamics underlies different functions of human KDM7 histone demethylases.* Chemistry - A European Journal, *25(21), 5422–5426.*

QM/MM optimized MD snapshots showed similar overall geometries but with fine and delicate differences, thus highlighting the effects of conformational flexibility on the enzyme-substrate complex (Chaturvedi et al., 2019).

5. Reactivity studies of model system

Density functional theory (DFT) studies on the demethylation of N,N,N-trimethyl, *N,N*-dimethyl- and *N*-methylpropan-1-aminium cations by [$(AcO)_2$(imidazole)$_2$(H_2O)Fe=O] showed that the methylation degree (number of methyl group attached to N) can affect the reaction mechanism (Chaturvedi et al., 2019; Alberro, Torrent-Sucarrat, Arrastia, Arrieta, & Cossío, 2017). For N,N,N-trimethyl- group bearing substrate, the mechanism is similar to other 2OG dependent enzymes while the probability of hydrogen abstraction from N generates alternative possibilities in *N,N*-dimethyl- and *N*-methylpropan-1-aminium cations (Alberro et al., 2017). For trimethylated residues, the mechanism begins with a hydrogen atom transfer (HAT) from one of the methyl group followed by the rebound step to form a hydroxymethylaminium intermediate (Scheme 2). However, the N-H hydrogen atom on *N,N*-dimethyl-, and *N*-methylpropan-1-aminium cations has comparable barriers as C-H hydrogen atom transfer step (Alberro et al., 2017).

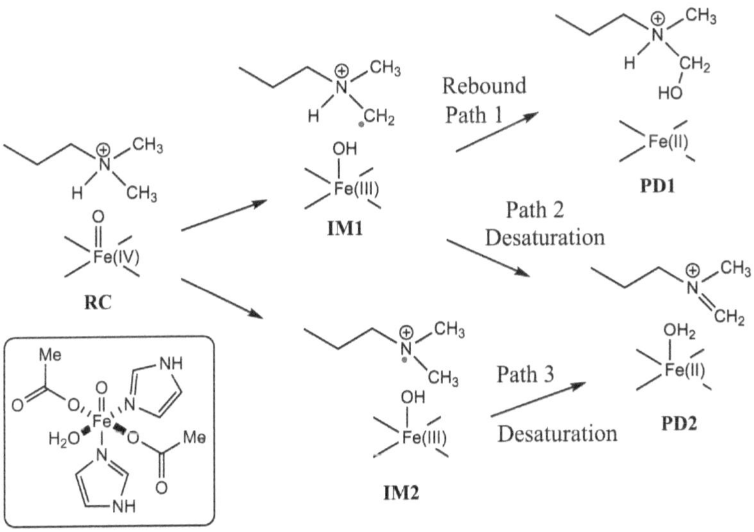

Scheme 2 Alternative hydrogen atom transfer (HAT) possibilities in di/mono methylated nitrogen in substrates.

6. Inhibitors of KDM7

Abnormalities in the activities of KDM7 enzymes result in cancer and neurological disorders (Copeland, Olhava, & Scott, 2010; Kelly, De Carvalho, & Jones, 2010). Most importantly, KDM7B is associated with proliferation of prostate cancer cells and osteosarcoma (Suzuki et al., 2013). Hence, the development of suitable inhibitors for the enzymes becomes important as they would aid in probing biological functions of the enzymes and to be used as anticancer drug candidate. Most inhibitors of the enzymes are iron-chelating 2OG competitors (Hopkinson et al., 2013; Rose et al., 2012) and have been reported to resemble the chemical structure of 2OG or succinate (Hopkinson et al., 2013; Rose et al., 2010). Many of these molecules possess high level of in vitro potency, but they are limited due to lack of activity and selectivity in cells (Gerken et al., 2017). Reduction in their activity in cells is due to high intercellular concentration of 2OG and poor cellular permeability (Gerken et al., 2017).

Several KDM2/7 inhibitors have been discovered and are presented in Fig. 5. Compound **1** was found to be the most potent and selective KDM2/7 inhibitor out of the series of compounds that were synthesized and studied by Suzuki et al. in 2013. Molecular docking informed about the importance of the cyclopropyl group of the compound as it formed strong interactions with the aromatic group of Phe 250 (that is present in the active site) through CH-π or cyclopropyl-π interactions. The strong interactions of the cyclopropyl group was linked to the small size of the ring as it fits well into the less spacious hydrophobic pocket of the enzyme (Suzuki et al., 2013). Daminozide (compound **2**) is a highly selective inhibitor of

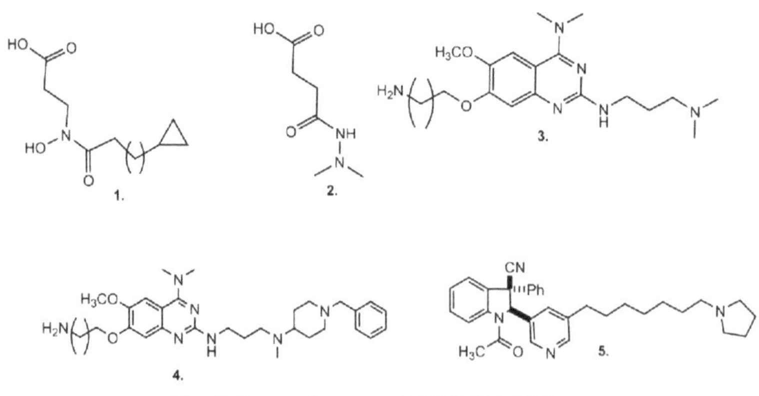

Fig. 5 Some discovered KDM2/7 inhibitors.

KDM2/7, and it was once used as a plant growth retardant but later withdrawn due to the carcinogenicity nature of its metabolite (1,1-dimethylhydrazine) (Rose et al., 2012). Compound **2** is at least 60-fold more selective as an inhibitor of the KDM2/7 subfamily over the other demethylases subfamilies as tested by Rose et al. in 2012. The inhibition by this compound works by binding in the 2OG binding pocket in a competitive manner and chelating to Fe via its acylhydrazide carbonyl and dimethylamino groups (Rose et al., 2012).

Compound **3** was generated from compound **4** by removing the benzylated six-membered piperidine ring. The crystallographic analysis of KIAA1718 JmjC domain in complex with compound **4** revealed that this moiety is solvent exposed and disordered, and thus its removal to give new compound **3** has no effect on the potency against PHF8 (Upadhyay et al., 2012). However, the removal leads to the reduction in the inhibitory effect against G9a-like protein (GLP) by a factor of 1500, implying that compound **3** is more selective against PHF8 than GLP by a factor of 35 (Upadhyay et al., 2012; Kelly et al., 2010).

Compound **5**, as reported by Gerken et al. in 2017, shows more than 75-fold selectivity toward KDM2A/7A when compared to other KDMs. Their study revealed that the compound possesses a different mode of inhibition to the majority of the previously discovered KDM inhibitors as it does not display competitive inhibition kinetic with respect to either 2OG co-substrate or the peptide substrate (Gerken et al., 2017).

7. Conclusions

Over the last several years, KDM7 enzymes were very intensively studied. The elucidation of the crystal structures of PHF8 and KIAA1718, together with biochemical studies, provided novel knowledge about their structure-function relationships. Many inhibitors were designed; however, their selectivity still remains a challenge. Modeling methods provided advanced insights which cannot be gained experimentally. There are however still open questions - what is the order of the substrate and co-factor binding, how dioxygen is transferred to the iron center and what is the reaction mechanism and the key interactions that stabilize the transition states and the reaction intermediates. Answering these questions would provide further important insights into KDM7's structure-function relationships and will help for the design of selective inhibitors as potential drugs.

References

Abidi, F. E., Miano, M. G., Murray, J. C., & Schwartz, C. E. (2007). A novel mutation in the PHF8 gene is associated with X-linked mental retardation with cleft lip/cleft palate. *Clinical Genetics, 72*(1), 19–22.

Aik, W. S., McDonough, M. A., Thalhammer, A., Chowdhury, R., & Schofield, C. J. (2012). Role of the jelly-roll fold in substrate binding by 2-oxoglutarate oxygenases. *Current Opinion in Structural Biology, 22*(6), 691–700.

Alberro, N., Torrent-Sucarrat, M., Arrastia, I., Arrieta, A., & Cossío, F. P. (2017). Two-state reactivity of histone demethylases containing Jumonji-C active sites: Different mechanisms for different methylation degrees. *Chemistry - A European Journal, 23*(1), 137–148.

Arteaga, M. F., Mikesch, J.-H., Qiu, J., Christensen, J., Helin, K., Kogan, S. C., et al. (2013). The histone demethylase PHF8 governs retinoic acid response in acute promyelocytic leukemia. *Cancer Cell, 23*(3), 376–389.

Bannister, A. J., & Kouzarides, T. (2004). Histone methylation: Recognizing the methyl mark. *Methods in Enzymology, 376*, 269–288.

Björkman, M., Östling, P., Härmä, V., Virtanen, J., Mpindi, J. P., Rantala, J., et al. (2012). Systematic knockdown of epigenetic enzymes identifies a novel histone demethylase PHF8 overexpressed in prostate cancer with an impact on cell proliferation, migration and invasion. *Oncogene, 31*(29), 3444–3456.

Chaturvedi, S. S., Ramanan, R., Waheed, S. O., Ainsley, J., Evison, M., Ames, J. M., et al. (2019). Conformational dynamics underlies different functions of human KDM7 histone demethylases. *Chemistry - A European Journal, 25*(21), 5422–5426.

Copeland, R. A., Olhava, E. J., & Scott, M. P. (2010). Targeting epigenetic enzymes for drug discovery. *Current Opinion in Chemical Biology, 14*, 505–510.

Gerken, P. A., Wolstenhulme, J. R., Tumber, A., Hatch, S. B., Zhang, Y., Muller, S., et al. (2017). Discovery of a highly selective cell-active inhibitor of the histone lysine demethylases KDM2/7. *Angewandte Chemie International Edition, 56*, 15555–15559.

Ghosh, A., Ghosh, S., Maiti, G. P., Mukherjee, S., Chakraborty, J., Roy, A., et al. (2012). Association of FANCC and PTCH1 with the development of early dysplastic lesions of the head and neck. *Annals of Surgical Oncology, 19*(Suppl. 3), S528–S538.

Hausinger, R. P. (2004). Fe(II)/α-Ketoglutarate-Dependent hydroxylases and related enzymes. *Critical Reviews in Biochemistry and Molecular Biology, 39*(1), 21–68.

Hopkinson, R. J., Tumber, A., Yapp, C., Chowdhury, R., Aik, W., Che, K., et al. (2013). 5-carboxy-8-hydroxyquinoline is a broad spectrum 2-oxoglutarate oxygenase inhibitor which causes iron translocation. *Chemical Science, 4*, 3110–3117.

Horton, J. R., Upadhyay, A. K., Hashimoto, H., Zhang, X., & Cheng, X. (2011). Structural basis for human PHF2 Jumonji domain interaction with metal ions. *Journal of Molecular Biology, 406*(1), 1–8.

Horton, J. R., Upadhyay, A. K., Qi, H. H., Zhang, X., Shi, Y., & Cheng, X. (2010). Enzymatic and structural insights for substrate specificity of a family of Jumonji histone lysine demethylases. *Nature Structural & Molecular Biology, 17*(1), 38–44.

Kelly, T. K., De Carvalho, D. D., & Jones, P. A. (2010). Epigenetic modifications as therapeutic targets. *Nature Biotechnology, 8*, 1069–1078.

Koivisto, A. M., Ala-Mello, S., Lemmelä, S., Komu, H. A., Rautio, J., & Järvelä, I. (2007). Screening of mutations in the PHF8 gene and identification of a novel mutation in a finnish family with XLMR and cleft lip/cleft palate. *Clinical Genetics, 72*(2), 145–149.

Laumonnier, F., Holbert, S., Ronce, N., Faravelli, F., Lenzner, S., Schwartz, C. E., et al. (2005). Mutations in PHF8 are associated with X linked mental retardation and cleft lip/cleft palate. *Journal of Medical Genetics, 42*(10), 780–786.

Lee, K. H., Park, J. W., Sung, H. S., Choi, Y. J., Kim, W. H., Lee, H. S., et al. (2015). PHF2 histone demethylase acts as a tumor suppressor in association with p53 in cancer. *Oncogene, 34*(22), 2897–2909.

Loenarz, C., Ge, W., Coleman, M. L., Rose, N. R., Cooper, C. D. O., Klose, R. J., et al. (2009). PHF8, a gene associated with cleft lip/palate and mental retardation, encodes for an Nε-dimethyl lysine demethylase. *Human Molecular Genetics, 19*(2), 217–222.

Luger, K., & Hansen, J. C. (2005). Nucleosome and chromatin fiber dynamics. *Current Opinion in Structural Biology, 15*(2), 188–196.

Ma, Q., Chen, Z., Jia, G., Lu, X., Xie, X., & Jin, W. (2015). The histone demethylase PHF8 promotes prostate cancer cell growth by activating the oncomir MiR-125b. *OncoTargets and Therapy, 8*, 1979–1988.

Rose, N. R., Woon, E. C., Kingham, G. L., King, O. N., Mecinovic, J., Clifton, I. J., et al. (2010). Selective inhibitors of the JMJD2 histone demethylases: Combined nondenaturing mass spectrometric screening and crystallographic approaches. *Journal of Medicinal Chemistry, 53*, 1810–1818.

Rose, N. R., Woon, E. C., Tumber, A., Walport, L. J., Chowdhury, R., Li, X. S., et al. (2012). Plant growth regulator daminozide is a selective inhibitor of human KDM2/7 histone demethylases. *Journal of Medicinal Chemistry, 55*, 6639–6643.

Shen, Y., Pan, X., & Zhao, H. (2014). The histone demethylase PHF8 is an oncogenic protein in human non-small cell lung cancer. *Biochemical and Biophysical Research Communications, 451*(1), 119–125.

Shi, Y., Lan, F., Matson, C., Mulligan, P., Whetstine, J. R., Cole, P. A., et al. (2004). Histone demethylation mediated by the nuclear amine oxidase homolog LSD1. *Cell, 119*(7), 941–953.

Sinha, S., Singh, R. K., Alam, N., Roy, A., Roychoudhury, S., & Panda, C. K. (2008). Alterations in candidate genes PHF2, FANCC, PTCH1 and XPA at chromosomal 9q22.3 region: Pathological significance in early- and late-onset breast carcinoma. *Molecular Cancer*, 7, 84.

Sun, X., Qiu, J. J., Zhu, S., Cao, B., Sun, L., Li, S., et al. (2013). Oncogenic features of PHF8 histone demethylase in esophageal squamous cell carcinoma. *PLoS One, 8*(10). e77353.

Suzuki, T., Ozasa, H., Itoh, Y., Zhan, P., Sawada, H., Mino, K., et al. (2013). Identification of the KDM2/7 histone lysine demethylase subfamily inhibitor and its antiproliferative activity. *Journal of Medicinal Chemistry, 56*, 7222–7231.

Upadhyay, A. K., Rotili, D., Han, J. W., Hu, R., Chang, Y., Labella, D., et al. (2012). An analog of BIX-01294 selectively inhibits a family of histone H3 lysine 9 Jumonji demethylases. *Journal of Molecular Biology, 416*, 319–327.

Walport, L. J., Hopkinson, R. J., & Schofield, C. J. (2012). Mechanisms of human histone and nucleic acid demethylases. *Current Opinion in Chemical Biology, 16*, 525–534.

Wang, Q., Ma, S., Song, N., Li, X., Liu, L., Yang, S., et al. (2016). Stabilization of histone demethylase PHF8 by USP7 promotes breast carcinogenesis. *Journal of Clinical Investigation, 126*(6), 2205–2220.

Yue, W. W., Hozjan, V., Ge, W., Loenarz, C., Cooper, C. D. O., Schofield, C. J., et al. (2010). Crystal structure of the PHF8 Jumonji domain, an Nε-methyl lysine demethylase. *FEBS Letters, 584*(4), 825–830.

Zhu, G., Liu, L., She, L., Tan, H., Wei, M., Chen, C., et al. (2015). Elevated expression of histone demethylase PHF8 associates with adverse prognosis in patients of laryngeal and hypopharyngeal squamous cell carcinoma. *Epigenomics*, 7(2), 143–153.

FSC
www.fsc.org
MIX
Papier | Fördert gute Waldnutzung
FSC® C083411

Kolibri 360 GmbH
Ferdinand-Jühlke-Straße 7
99095 Erfurt, Deutschland
produktsicherheit@kolibri360.de